BEAR NECESSITY

A Novel

JAMES GOULD-BOURN

SCRIBNER

New York   London   Toronto   Sydney   New Delhi

Scribner
An Imprint of Simon & Schuster, Inc.
1230 Avenue of the Americas
New York, NY 10020

Originally published in Great Britain in 2020 by Trapeze Books

This Scribner trade paperback edition August 2020

SCRIBNER and design are registered trademarks of The Gale Group, Inc., used under license by Simon & Schuster, Inc., the publisher of this work.

For information about special discounts for bulk purchases, please contact Simon & Schuster Special Sales at 1-866-506-1949 or business@simonandschuster.com.

The Simon & Schuster Speakers Bureau can bring authors to your live event. For more information or to book an event, contact the Simon & Schuster Speakers Bureau at 1-866-248-3049 or visit our website at www.simonspeakers.com.

Interior design by Kyle Kabel

Manufactured in the United States of America

1   3   5   7   9   10   8   6   4   2

ISBN 978-1-9821-5288-8
ISBN 978-1-9821-2831-9 (ebook)

For my mum and dad,
Linda and Phillip Gould-Bourn

CHAPTER 1

Danny Malooley was four years old when he learned the hard way that lemon-scented soap tasted nothing like lemons and everything like soap. When he was twelve, while saving a cat that may or may not have needed saving, Danny learned the hard way that there was no such thing as a painless, nor dignified, way to fall out of a sycamore tree. When he was seventeen, he learned the hard way that all it took to become a father was a three-liter bottle of cheap cider, a girlfriend to share it with, an awkward fumble on the Hackney Downs, and a general disregard for the basic laws of nature; and when he was twenty-eight, he learned in the hardest way imaginable that all it took to dim the stars, stop the clocks, and bring the earth to a shuddering halt was one small, invisible sliver of ice on a country road.

A screech of tires tore Danny from his sleep, or it could have been a scream, he wasn't quite sure. He sat up and scanned the room, trying to connect the sound with his surroundings until his brain woke up and told him it was a nightmare. Lying back down on his sweat-soaked pillow, he looked at the clock on the bedside table: 6:59 a.m., the digits bright in the morning gloom. He switched off the alarm before the numbers rolled over and gently ran his hand across the empty pillow beside him. Then, heaving the clammy duvet aside and crawling out of bed, he ignored his reflection in the wardrobe mirror and slowly dressed in yesterday's clothes.

Will's bedroom door was ajar, so Danny pulled it shut on his way to the kitchen. Filling the kettle and setting it to boil, he dropped some dry but not yet furry bread into the toaster and turned on the radio, more out of habit than a desire to know what was happening in the world. The newsreader murmured to herself in the background while he surveyed the postcard view from the window—"postcard" due to the size of the window, not because of the beauty beyond it. The sky was as blue as the Victoria Line, but the beaming sun did little to brighten the landscape. Danny often thought the housing estate actually looked worse in the sunlight, mainly because more of it was visible. Just as poor lighting could make a Tinder date attractive or a run-down restaurant quaint, so too could a leaden sky help to partially conceal the full grim reality of the Palmerston estate. As he gazed at the wall of concrete housing blocks that mercifully obscured his view of even more concrete housing blocks, Danny once again resolved to move, just as he had done yesterday, and just like he'd do again tomorrow.

He ate his breakfast at the dining room table, his eyes fixed on the same wall he'd stared at so much over the last fourteen months that the paper had started to curl beneath the weight of his gaze, but Danny hadn't noticed. Nor had he noticed the darkening patch of carpet in the hallway, sullied by the work boots he kicked off every day without first banging the mud from their soles, or the film of grime on the windows that gave whoever looked through them an early glimpse of what to expect from cataracts, or the potted carcass on the windowsill that had once been a healthy philodendron but now resembled a clump of irradiated potato peel. He wouldn't even have noticed the post were it not for the fact that it always arrived during breakfast, causing him to flinch as it clattered through the letterbox and landed on the mat.

Two white envelopes sat in the hallway. The first contained a passive-aggressive reminder from his water provider that he was two months behind on his payments. The second was a final notification about his unpaid electricity bill, much of it written in bold red letters, especially the words *court, bailiff, prosecution*, and, somewhat bizarrely, *thank you*, which made it seem more like a threat than a common expression of gratitude.

Danny frowned and stroked his stubble, the four-day bristles rasping beneath his nail-bitten fingers. He looked at the whiteboard on the wall where a thick wad of paper was held in place by a couple of souvenir magnets from Australia. Above it, written in bold black letters, was the word *UNPAID*. Two sheets of paper hung next to the bundle. This was the *PAID* pile. He added the new arrivals to the bigger stack, which held for less than a second before the magnets gave way and dumped the bills in a fluttering mess across the floor. Danny sighed and gathered them up. Then, using a third magnet, this one shaped like the Sydney Opera House, he reattached the bills to the whiteboard and scribbled *Buy more magnets!* beside them.

"Will!" he shouted from the kitchen doorway. "You up?"

Will heard his dad but didn't respond as he continued to examine the bruise on his arm. It looked like a storm was raging between his bony shoulder and what passed for his bicep, a blue-black cloud on milky-white skin. Will gently probed it with his finger, unaware of just how tender it was until the slightest pressure triggered a dull ache that seemed to engulf his entire upper arm.

"Come on, Will, breakfast!" shouted Danny, his voice already weary.

Will plucked his crumpled school shirt from the door handle and winced as he carefully fed his arm through the sleeve.

"Morning, sleepyhead," said Danny as Will shuffled past the kitchen door and slumped down at the table. Danny joined him a few minutes later with a mug in one hand and a plate of toast in the other. He put them down in front of Will and took the seat opposite.

Will studied the plate through his sandy-blond fringe, which covered the two-inch scar at his hairline. Thomas the Tank Engine peered at him between two slices of peanut-buttered toast while James the Red Engine grinned almost mockingly from the mug.

"Eat up or you'll be late," said Danny. He took a mouthful of cold tea and grimaced.

Will swiveled his mug until the train disappeared from view. He took a tentative bite of his toast and placed the remainder over Thomas's face.

"Remember it's your mum's birthday today," said Danny.

Will stopped chewing and stared at his plate. The murmur of the radio crept into the silence between them.

"Will?" said Danny.

Will nodded once without looking up.

The doorbell rang and Danny stood to answer it. He squinted through the spyhole to find Mohammed waiting in the open-air corridor. The boy was chubby with thick-rimmed glasses and a hearing aid behind each of his ears. London lurked over his shoulder.

"Hi, Mr. Malooley," he said as Danny opened the door. "Did you know that a blue whale's fart bubbles are so big you can fit an entire horse inside them?"

"No, Mo. I can honestly say I did not know that."

"Saw it on Animal Planet last night," said Mo, who enjoyed watching wildlife documentaries as much as most eleven-year-olds enjoyed watching people seriously injure themselves on YouTube.

"Sounds a bit cruel," said Danny. "How did they even get a horse inside a whale fart?"

"Don't know," said Mo. "They didn't show that bit."

"Right." Danny frowned as he pondered the logistics of such an experiment.

"Is Will ready yet?"

"Give him two mins, he's just eating—"

Will barged past Danny and into the corridor before he could finish his sentence.

"Bye, Mr. Malooley," said Mo as Will roughly guided his friend towards the stairwell.

"Bye, Mo. Will, see you after school, okay?"

Will didn't respond as he disappeared around the corner.

Back in the living room, Danny gathered the cups and plates from the table. He poured Will's untouched tea down the sink and tipped his uneaten toast into the bin. It was the same routine he'd performed almost every day since the accident.

CHAPTER 2

Danny crossed the building site in a yellow hard hat and a high-vis jacket that flapped in the wind. He aimed for Alf, the foreman, who was similarly dressed but holding a clipboard. Alf was a stout and balding man with a face like a boxer who never kept his guard up. Noticing Danny approaching, he looked over his shoulder at the black-suited, bony-faced, pale-skinned man standing nearby, who could have been mistaken for Death were he not wearing a safety helmet. The man tapped his watch and pointed at Danny. Alf sighed.

"Morning, Alf," said Danny, shouting over the noise as cranes loaded with pallets pivoted slowly overhead while the shuddering arms of excavators scooped up huge wads of earth.

"You're late, Dan."

Danny frowned and checked his phone. "Not by my clock," he said, showing the screen to Alf.

"By his," said Alf, ignoring the phone and nodding towards the man in the suit.

"Who's that?" said Danny.

"Viktor Orlov. New project manager."

"Orlov?"

"Cossack," said Alf. "Real ball-breaker. Already fired two people this morning. He's coming down hard on everyone."

Danny stared at the man in the suit. The man stared back with a frosty gaze.

"Anyway, get moving," said Alf. "You're on cement with Ivan. And, Danny?"

"Yes, Alf?"

"Don't be late again."

Danny grabbed a shovel and went to join Ivan, a Ukrainian man-mountain of muscle and broken English who could move more earth than an excavator and build things quicker than a Minecraft champion. Danny suspected that Ivan had killed at least one person in his lifetime, probably with his bare hands. This hunch was largely inspired by the gallery of crude prison tattoos that adorned his bulging forearms, which were covered with jagged words, ugly faces—there was even a completed noughts-and-crosses board on his left arm, near the elbow—and other random scribbles that Danny was too afraid to ask about.

The two had been friends since Danny saved Ivan's life a couple of years ago. That, at least, was how Danny and everybody else on the building site remembered it, but Ivan refuted this version of events. Ivan had only been on the job for two weeks when a rogue piece of scaffolding came loose in a gale. The steel tube would have landed directly on his head had Danny, who happened to be working nearby, not barged the big man out of the way (almost dislocating his shoulder in the process). But while Danny was hailed a hero that day, Ivan, who had, in his own words, once been run over by a tank and survived, stubbornly maintained that a thirty-kilo pole to the head was unlikely to even cost him a sick day, let alone kill him, and that everybody was just being melodramatic "like the *EastEnders*." The whole thing had become something of a running joke between them, although Danny was the only one who seemed to find it funny.

"Danylo," said Ivan as he slapped a wad of cement into a wheelbarrow.

"All right, Ivan. Who's the tool in the suit?" Danny cast a thumb over his shoulder.

"So," said Ivan, "you have met Viktor."

"Alf says he's already fired two people this morning."

"They send him from Moscow. They say we do not work fast enough."

"And they think we'll work faster if they fire us?" said Danny.

Ivan shrugged. "In Ukraine we have word for man like Viktor."

"Oh yeah?" said Danny. "What?"

"Asshole," said Ivan.

Danny laughed. "How was your holiday?" he asked, digging into the wet cement.

"Holiday?" said Ivan. "What holiday? I take Ivana to Odessa. I spend the week with her family. Her mother, she hate me. And her father. And her sister. Even the dog hate me."

"I can see," said Danny, pointing to a set of teeth marks on Ivan's forearm.

"What?" said Ivan, following his finger. "Oh. No. That was her grandmother."

"Right."

Ivan removed a bundle of paper from his pocket and sheepishly handed it to Danny.

"Here," he said.

Danny knew what it was before he'd even opened it. A week after the scaffolding incident, Ivan had invited Danny, his wife Liz, and Will over for dinner. They'd barely spoken since Danny had (or hadn't, depending which camp you were in) stopped Ivan from getting skewered by a six-foot pole. Apart from that day, in fact, the two men had barely spoken *at all*, and Ivan gave no explanation for the invitation, although Danny had always taken it to be a subtle form of thank-you. They'd spent what turned out to be the first of many evenings around a dinner table together, eating, laughing, and drinking too much *horilka* (Liz drank more than anybody, and consequently suffered more than anybody) while Will and Yuri—Ivan and Ivana's son—played Xbox and bonded over their mutual embarrassment at seeing their parents having fun. At some point in the night, Liz had fallen in love with Ivana's collection of painted wooden eggs that she kept on the windowsill. Ever since then, whenever they went back to Ukraine, Ivan had returned with a wooden egg for Liz, something he'd continued to do despite the tragic change in circumstances.

"Thanks," said Danny, turning the colorful ornament over in his hand. He knew how awkward these moments were for Ivan, who must have wondered more than once whether or not to abandon the tradition, but Danny was grateful that he hadn't.

"How is Will?" said Ivan, keen to move the conversation along.

"He's fine," said Danny, slipping the egg into his jacket pocket. "I guess. I don't know."

"He still does not talk?"

"Nope. Not a word. Not even in his sleep."

A man arrived with an empty wheelbarrow and waddled away with the full one.

"You know," said Ivan, "maybe he *is* speaking."

"Not to me, he isn't."

"No, I mean, being quiet can also be loud, you understand?"

"Not really, no," said Danny.

"Look," said Ivan, standing his shovel in the wet cement and leaning on the handle. "When Ivana she is angry with me, sometimes she yell and call me stupid asshole, but sometimes, when she is really angry, she say nothing for many days. She is quiet, like mouse, but I know she is still telling me something, you know?"

"Like what?" said Danny.

Ivan shrugged. "Like how she would like to put my head in oven."

"You think Will's trying to tell me to put my head in the oven?"

"No, but maybe you just do not hear what he is saying."

"Well, if he's trying to tell me something, I wish he'd just come out and say it," said Danny. "It's been over a year now. Whatever he wants to say to me, it can't be any worse than the silence."

CHAPTER 3

Girls pretended not to watch boys while gossiping in groups or playing with their phones, and boys pretended not to watch girls while secretly trying to impress them, mainly by playing keep-away and filming each other thumping unsuspecting classmates in their nonvital organs. Everybody was watching everybody, but nobody made eye contact. It was like one big staring competition, one where you could blink as much as you wanted to but shriveled up like a salted slug if someone caught you looking at them. Only one person had the confidence to hold the gaze of every pair of eyes in the schoolyard, and that day, like most days, Mark had Will in his sights.

"Seriously, it was crazy!" said Mo as they weaved their way through the crowds towards school. "These lions, there were, like, eight of them or something, well, lionesses actually, lions don't really hunt, and they were eating this buffalo, or a bison or whatever, but it was still alive, and it was just, like, standing there and eating grass while they were eating *it*, and—"

Will jabbed Mo in the ribs with his elbow.

"What was that for?" said Mo, rubbing his side.

Will nodded at the three scruffy boys approaching from across the yard. They were taller and older than Will and Mo, and they swaggered like they knew it. Their shirts were untucked and their ties were loose like a trio of overworked detectives, but if Mark and his goons were spotted near a crime then they probably weren't trying to solve it. Mark was the shortest member of his posse by a good few inches, but the boy made up for his lack of stature, and looks, and intelligence, with his reputation as Richmond High's most notorious terrorist. You didn't have to do anything wrong to find yourself on his bad side (otherwise known as his *only* side). Simply existing was enough to have your name involuntarily entered into the Markus Robson lottery of pain, and for reasons that Will had never been able to fathom, his name seemed to come up at least twice as often as anybody else's.

"Come on," said Mo. They picked up the pace, suddenly eager to get to class. The older boys also sped up, scurrying through the crowd like three ferrets after the same trouser leg.

"Look who it is, lads," said Mark as he blocked the main entrance. "Dumb and dumber. Or should that be *deaf* and dumber?"

"I told you already, I'm not deaf," said Mo. "I have—"

"What?" asked Mark, cupping his hand behind his bigger ear. "I can't hear you, mate."

"I said I'm not deaf, I just—"

"What?"

"I said I'm—"

"Can't hear you, Mo, speak up," said Mark.

Mo sighed, the joke finally sinking in. "Idiot," he muttered as he fiddled with his hearing aid.

"What was that?" said Mark.

"I thought you couldn't hear me?" said Mo sarcastically.

"Best watch that mouth of yours, Mo," said Mark. He yanked Mo's tie with a violent tug that turned the knot into a peanut. "Learn a trick from your boyfriend here."

Mark turned on Will while Mo struggled to loosen his tie.

"What you looking at?" he said.

Will shrugged and stared at his shoes.

"You fancy me?" said Mark. "Is that it?"

Will shook his head.

"So you're saying I'm ugly?"

Another shake of the head.

"So you fancy me, then?" said Mark.

"Leave him alone," said Mo.

"Shut it, Mo-by dick-head," said Mark.

"Moby dickhead," said Tony, the taller of Mark's two goons. "Good one."

"I don't get it," said Gavin, who had so many zits that his head contained more pus than brains.

"Moby Dick," said Tony. "You know, like the book. With the whale and the one-legged Arab and whatever."

"Arab?" said Gavin. "Like Mo?"

"It's Ahab," said Mo. "Captain Ahab. And I'm not Arab, I'm Punjabi."

"Same thing," said Gavin.

"*Teri maa ka lora*," muttered Mo.

"How's your arm?" said Mark, pointing at Will's bicep.

Will shrugged with as much false bravado as he could muster under the circumstances, which wasn't very much at all.

"Won't mind if I deck you again, then, will you?" said Mark. He feigned a punch, and Will's hand instinctively moved to shield his arm. Mark grinned. "Thought as much," he said. The school bell rang and they turned to leave. "See you at lunch, losers."

Mo rubbed his neck and quietly cussed them again in Punjabi. Will nodded, sure that whatever Mo had said was bad.

They joined the other students who were filtering into the building and made their way to class. Taking a seat at his desk beside Will's, Mo nudged his friend and pointed at the thin-haired,

Brillo-bearded man in glasses who was standing with his back to the whiteboard. He looked like he'd dressed in the dark and he wore an expression like he didn't particularly care.

"Where did this guy escape from?" said Mo. Will shrugged.

"Okay, everybody, settle down," said the man, his voice imbued with the weariness of someone who spent his entire life being ignored. "You're probably wondering who I am and why I'm here. And, to be honest with you, I sometimes ask myself those same questions, as will each and every one of you in this room one day when you realize that life is nothing but one long series of disappointments. But just to clarify, my name is Mr. Coleman and I am your substitute teacher."

He scribbled his name across the whiteboard and underlined it.

"Not Cullman. Not Collman. Not Cool Man, although feel free to call me that. Otherwise it's Mr. *Coleman*. Got that?"

A murmur of acknowledgment rose from the class.

"I'll take that as a yes. Now, before any of you make the grave mistake of thinking I'm an easy target because I'm new, think again. I have seen and heard just about everything that can be seen and heard in a classroom, so whatever you did to scare off Mr. Hale, rest assured that it won't work with me. Do I make myself absolutely clear?"

Mr. Coleman eyeballed the class, extinguishing every smile he came across.

"Great. Now, let's start with the attendance, shall we? It's a simple enough process. I call your name and you shout, 'Present.'"

Mr. Coleman opened the register and briefly flicked through the pages.

"Atkins?" he said, his pen hovering above the page.

"Present," said a girl with braces who sat in front of Will.

"Well done, Sandra," said Mr. Coleman as he dashed off a tick beside the girl's name. "You've clearly done this before. Cartwright?"

"Here," said a boy with a squiffy tie who sat at the back of the class.

"Unlike Cartwright, it seems," said Mr. Coleman. Everybody laughed but Cartwright. "Jindal?"

"Present," said Jindal.

"Take note, Cartwright," said Mr. Coleman.

"Present," said Cartwright to the sound of more laughter.

"No, Cartwright, I've already . . . forget it. Kabiga?"

"Present," said Kabiga.

"Malooley?"

Silence.

"Malooley?"

A few sniggers punctuated the quiet as Mr. Coleman scanned the room. All the desks were occupied. Will sat with his hand in the air. Mr. Coleman frowned.

"Yes?" he said.

"He's Malooley, sir," said Mo.

"Is he?" said Mr. Coleman. He looked at Will. "Then why didn't you say 'Present'?"

"He doesn't speak, sir," said Mo.

"He . . . doesn't speak?"

"No, sir."

"And you're, like . . . what? His representative?"

"More like his spokesperson, sir," said Mo. A ripple of laughter passed through the class.

"Right," said Mr. Coleman. He dropped his eyes to the register and drew a tick beside Will's name. "I take it back. *Now* I've seen everything."

Will spent the first part of his lunch break in the caretaker's cupboard. He often spent some part of his school day in there, not because he enjoyed the smell of industrial cleaning products or the sensation of sitting in a darkened room for prolonged periods of time, but because Mark and his gang had once again locked him in there after ambushing him on his way to the cafeteria. This had been happening since the day they'd discovered, while up to no good, that the inside handle of the cupboard door was loose and could be removed with very little effort, thereby creating a makeshift holding cell for their hapless victims that could only be opened from the outside. Will had the inauspicious honor of being their very first inmate. He was also the longest-serving, having once been trapped in there for two whole periods, although given that those periods were maths and science, he didn't exactly go to great efforts to liberate himself.

He actually quite enjoyed the silence and the solitude of the cupboard these days. He didn't even put up a fight when they locked him in there anymore (which ruined their fun slightly, but not enough to stop them from doing it). Nobody could laugh at him or mock him or insult him in there. Nobody could call him an attention-seeker (something that Will found particularly annoying given how much effort he put into *not* being noticed), and nobody could beat him up because the people who usually did the beating were the same people who had locked him in the cupboard to begin with. Also, nobody was pretending to know how he felt. Nobody was comparing his situation to their own because they'd once had a sore throat or lost their voice for a week. Everybody just left him alone. The only downside to the arrangement was that he got hungry, so when Mo texted to find out where he was, Will was about to text back when he heard Mrs. Thorpe's voice in the hallway.

"Oh, hi, Dave."

Sue Thorpe was the head teacher. Unlike many heads of school, however, slate-faced disciplinarians with nose hair longer than their tempers and an inability to look at a ruler without wanting to whack somebody with it, regardless of whether they were a student or not, Mrs. Thorpe was funny, personable, and generally well-liked by the students, even if she had to sometimes suppress the urge to assault them with stationery.

"Sue, good to see you." It took Will a second to recognize Mr. Coleman's voice.

"How was it this morning?" she said.

He heard Mr. Coleman sigh. "Well, you know that feeling when you look around the classroom and everybody is listening to what you're saying and you can almost *see* them getting smarter, and you stand there and think to yourself: *This is why I became a teacher. This is what it's all about?*"

Mrs. Thorpe paused for a moment. "Not really," she said.

"Exactly," he said. Will smiled.

"Business as usual, then?"

"Business as usual," said Mr. Coleman. "Actually, no, that's a lie."

"Oh? Do tell."

"What do you know about a boy called Malooley?"

"Will?" she said.

"Yeah," said Mr. Coleman. "The quiet one."

Will shuffled over to the door and pressed his ear against it.

"He's a nice kid. Good student. Why do you ask?"

"Can he really not talk? Or is this just part of my new teacher initiation ceremony?"

"He *can* talk," said Mrs. Thorpe. "He just, well, doesn't want to. Selective mutism, they call it."

"Wow. I wish my kids had some of that."

"Tell me about it."

"Has he always been like that?" said Mr. Coleman. Will was painfully aware of what Mrs. Thorpe was going to say next.

"His mum died about a year ago. Car crash. She hit an icy corner and went straight into a tree. Will was in the car at the time, poor kid. He hasn't spoken since."

Mr. Coleman muttered something that Will didn't catch but presumed to be an expletive. Whatever it was, Mrs. Thorpe agreed.

"He gets a bit bullied about it by the older boys, so keep an eye out. I've had a few words with them, but you know what teenagers are like."

"Sadly."

Their voices grew fainter as they walked off together down the corridor.

Will stayed in the cupboard for another few minutes, his appetite suddenly gone, but the room felt darker than it did before, so he texted Mo to come and let him out.

CHAPTER 4

The school bell rang and a flood of children poured from the entrance and across the yard. Danny scanned the sea of red uniforms for Will until he and Mo emerged with Mark and his goons on their heels. Gavin was throwing peanuts at Mo and Tony was repeatedly treading on the back of Will's shoe, causing him to stumble. Mark walked behind them, proudly grinning at his well-trained underlings until, noticing Danny glaring at them, he grabbed his mates and faded into the crowd.

Will waved good-bye to Mo and slowly crossed the road with his head down and his hands in his pockets.

"Who are they?" asked Danny, nodding towards Mark.

Will shrugged and shook his head.

"I'd sue my parents if I looked like that."

Will cracked a smile that was more like a prelude to a proper smile that never arrived.

"You'd tell me if they were giving you a hard time, wouldn't you?"

Will nodded. Danny looked unconvinced.

"Come on," he said.

*   *   *

Will stared at the ground while his dad scanned the epitaphs, all of which looked cold and dull beneath the pigeon-wing clouds that had gathered overhead.

Danny knew precisely where they were going, but he still took his time, not because he wanted to be there—he didn't, and he knew Will didn't either—but because despite more than a year having passed since the accident, he hadn't yet processed his grief to the point where he could fully accept that his wife was dead, at least not in the conventional sense of the word. He knew she was gone. That much he understood. What he *couldn't* understand was the idea she was gone *forever*. Instead he imagined her gone in the same way his father was gone: not dead (or so he assumed, although he really had no idea, nor did he care to know), but not present either. It was, in some ways, an even crueler concept of death than death itself, because it did what death could not, which was to give him hope—no matter how small—that he might one day walk around a corner or through a door and find his wife standing on the other side of it. Sometimes he was sure he could smell her perfume in a room he'd just entered, or hear her voice on a crowded street, or feel her hand against his face as he roamed the lonely periphery of sleep. Other times she felt so close to him that all he had to do was turn around, but she'd be gone by the time he looked over his shoulder, her body swallowed by the crowd, her voice carried off by the wind. It was as if she occupied a world that ran parallel to his own, like two strangers living in a high-rise who could hear each other's movements but never crossed paths, which was why he was always so reluctant to visit the cemetery. Nothing destroyed that illusion more than seeing his wife's name etched into a cold and lifeless slab of granite.

"Here we are," said Danny, pausing beside a black headstone with gold lettering. He crouched and placed his hand on the stone while Will hovered nearby.

The grave was a simple arrangement, the small plot a far cry from the elaborate statues and monuments that stood in mournful silence around them. A rectangular border enclosed a layer of shiny green glass chippings that caught the light when the sun was out and sparkled like the surface of a lake. Today, however, they looked as drab as the last bunch of flowers that Danny had brought, their brown stems wilting from the pepper-pot lid of the metal vase.

"Always liked tulips, didn't she?" he said, dragging the old flowers from the pot and replacing them with new ones. He carefully arranged them and wiped a fleck of imaginary dirt from the headstone. "Think she'll like the color?" he asked, turning to look at Will. "They didn't have any yellow ones left."

Will stared at the grave, his jawbone tight.

"Want to say something to your mum?" said Danny. "For her birthday?"

Will shook his head, his eyes fixed on his mum's name etched into the stone.

"Go on," said Danny, putting a hand on his shoulder. "Have a go."

Will shrugged out from beneath his hand and marched off down the path.

"Will!" shouted Danny before sheepishly apologizing to an old lady who scowled at him from a nearby graveside. He watched Will take a seat on a bench at the far end of the cemetery.

"He's getting more like you every day, Liz," he said. "Seriously, I don't know what to do with him. I've tried everything, but he just won't talk. He barely even looks at me half the time. I don't know if he loves me or hates me or what. I keep hoping he'll grow out of it, like this is just a phase or something, but the longer it goes on, the more it feels like this is forever, whatever *this* even is." He sighed and shook his head. "Sometimes it feels like I lost both of you that day."

The leaves hissed in the branches above him as the trees gently creaked in the wind.

"Sorry, Liz," said Danny. He blinked a few times and exhaled like he'd just emerged from ice-cold water. "Right life of the party I am. We're fine. Everything's fine. Well, not fine, but, you know, we're getting there. Will's doing well in school, work is still work, our landlord is still a wanker, and Mrs. Amadi from flat thirty-six still thinks your name is Susan. She also thinks that Will isn't talking because evil spirits stole his voice, so she kindly gave me the telephone number of a nice man called Alan who performs very reasonably priced exorcisms, apparently. So, yeah, there's that."

He laughed, or tried to, but the sound that emerged was as empty as it felt.

"Listen to me," he said, glancing at the overcast sky. "I'm standing here, talking to a stone, and I know you can't hear me because you're not here. You *can't* be here because the sun isn't shining, which means I am literally talking to a rock right now while you're out celebrating your birthday without me. So I'll leave you to it, beautiful. Wherever you are, and whatever you're doing, I hope you're smiling, and I hope you're dancing. Just try not to wake me up when you get home, okay?"

Danny touched his lips and placed his fingers on the headstone.

"Love you, Liz. Happy birthday."

CHAPTER 5

They bought some chips and ate them in the park. Neither of them was hungry and Danny stabbed at his food with disinterest, while Will flicked it from his tray for the pigeons to eat. Several street performers were entertaining people nearby, singing, dancing, and doing whatever else it took to charm bystanders into opening their wallets. One scruffy man with long, matted hair and a tatty panama hat was strumming a guitar. It wasn't his music that drew the crowds but the portly beige cat in the knitted red sweater that was sitting on the man's shoulder and meowing at random intervals. Another man in a purple robe and a matching pointy hat was busy performing magic tricks, his face set into a serious frown as he wiggled his fingers at things and uttered seemingly ancient incantations. A smaller crowd had gathered around somebody dressed as a giant squirrel who was juggling football-size hazelnuts, and another person in a chicken costume was trying and failing to get people's attention by trying and failing to break-dance.

As Danny watched the various acts he couldn't help but notice how much money the performers were making. Their upturned hats, felt-lined instrument cases, Tupperware containers, and scuffed tobacco tins were literally overflowing with coins. Even the dancing chicken had somehow persuaded people to cough up their hard-earned cash, and all he was doing was writhing around as if hornets had set up home in his underpants.

Danny skewered a chip with his fork and gently nudged Will with his elbow.

"I think I'm in the wrong job," he said.

The sun was setting by the time they got home.

"You got any homework?" said Danny as Will emerged from the bathroom, his wet hair flat against his head and a smear of toothpaste along one cheek.

Will shook his head.

"Want to watch some TV or something?" asked Danny, already knowing the answer.

Will faked a yawn and pointed to his room.

"All right, well, lights out by nine, okay?"

He nodded and opened his door.

"Will," said Danny. His son paused but didn't turn around. "I know it's hard, but it'll get easier. I promise. It just, you know. It takes time."

Will looked at Danny, who gave what he hoped was an encouraging smile. Neither of them seemed convinced. He nodded once and closed the door behind him.

Danny turned on the television to commence his nightly ritual of sitting alone in front of the box until the early hours of the morning. His eyes felt heavy and his body felt tired, but he knew that any attempt to sleep would result in a long night of staring at the ceiling or watching the clock as the minutes turned into hours and the hours turned into daylight. Even on the rare occasion that he managed to get a proper night's rest, Danny often felt worse than if he hadn't slept at all, because waking up to confront the day meant also having to confront the fact that Liz wasn't beside him.

He thought about their last morning together when he'd woken up to find that Liz, as per usual, had commandeered the duvet at some point during the night. She always denied doing it, but every morning, without fail, he'd wake to find her fast asleep in the scene of her very own crime, bundled up in the covers while he lay there shivering in his underpants. That morning, however, after shuffling up to the ball of duvet and snuggling close to his wife—a gesture partly born from affection but mainly from a desire to keep warm—Danny was startled when the covers deflated beneath the weight of his arm. It wasn't until he heard Liz humming along to the kitchen radio that he realized she wasn't in bed beside him. He'd laughed back then, but it felt like a hideous joke whenever he thought about it now, as if some cruel higher power were preparing him for a life without Liz by dropping a clue so spitefully subtle that he had no way of solving it before his wife climbed behind the wheel and gave him what neither of them knew was a final kiss good-bye.

She wasn't even supposed to be driving that day, and the fact that she had had caused a rift between Danny and Roger, his father-in-law, which had never been resolved. It wasn't so much a new rift as a widening of an old one that had been expanding ever since Liz had first brought Danny home and introduced him as the man she was going to marry one day, a pronouncement that took her parents by surprise, not least because she was only sixteen and they didn't even know she had a boyfriend until then. Danny hadn't wanted to go, certain her parents would hate him, particularly her father, who happened to be a policeman and was therefore suspicious of everybody, especially teenage boys, and especially a teenage boy from Newham with a father who'd been so absent that nobody even noticed when he walked out on his son's fourteenth birthday, and a mother who kicked him out when her new boyfriend decided that the flat wasn't big enough for the three of them.

Still, Liz insisted it would all be fine, so Danny had reluctantly complied. After meeting her mother, Carol, who gave him a warmer hug than his own mother ever had, he started to believe her. Only when he met Roger did he realize Liz had either flat-out lied or grossly underestimated her father's temperament, because although the man shook Danny's hand in a feigned display of cordiality, the bone-crunching pressure he applied to that handshake told Danny everything he needed to know about the man's feelings towards him. It wasn't a handshake to assert authority. Nor was it a test of masculinity. It was a handshake that told him, in no uncertain terms, that Roger would rather be squeezing his neck than his hand and would do just that if he ever got the chance.

Danny knew that the man believed he was leading his daughter astray, something he always found slightly unjust considering that Liz, perhaps in defiance of her law-abiding upbringing, was often the more rebellious of the two. But Roger never saw that side of her, being hardwired, like most fathers, to see nothing but the good in his daughter, even when faced with overwhelming evidence to the contrary. His instinct was to blame Danny for anything that chipped away at the innocent and infallible image of the daughter that he carried in his wallet. He blamed him when Liz gave up the ballet lessons she'd been attending since she was six, even though Danny had actively encouraged her to continue with them. He blamed him when Liz became pregnant, which Danny couldn't exactly dispute but nevertheless found a little unfair given that Liz had been the primary instigator that fateful night on the Downs (he kindly spared Roger this information). And most crushingly of all, he blamed Danny for the death of his only child, something Danny knew not because of some wild inkling or nagging suspicion but because the man had said precisely that at Liz's funeral. To the eternal embarrassment of Carol, who had always been good to Danny, and to the shock of everybody else at the reception, Roger had told him that his daughter would still be alive if only he'd been the one in the driver's seat that day. It wasn't a backhanded compliment about Danny's driving abilities (the man never gave him compliments, backhanded or otherwise) but a reference to how Liz didn't really like driving, even though Roger was the one who had bought her a car in the first place. If Danny had been behind the wheel, he might have taken that icy corner at a slightly different speed, or at a slightly different time, or at a slightly different angle. Even if all of those alternative scenarios had still resulted in the same twisted wreckage by the roadside, at least they might have buried Danny and not Roger's beloved Elizabeth.

As painful as it was to hear, and as poorly timed as Roger's outburst was, Danny knew that the man had a point. Not a day had passed since the accident that he hadn't thought about how different things might have been if he'd called in sick that day, or if he'd only held on to her for a few more seconds before letting her get into the car, or if he'd left his work boots in the hallway again, forcing Liz to delay her trip while she wearily reminded him of the house rules in that voice he used to find so annoying but would now gladly trade his right arm to hear again.

Danny might even have forgiven Roger had he chosen to end things there, accepting his monologue as nothing more than the desperate words of a grieving father who was simply trying to make sense of something that could never be understood, but Danny couldn't forgive the hate with which the man had spat his concluding words.

"And now he's stuck without a mother," Roger had said, pointing at Will, who had only recently been discharged from the hospital and still wore a bandage around his head that shone like a beacon amidst the black shirts and dresses. "Now he's stuck with *you*."

Danny had stayed quiet until then, determined not to make even more of a scene, but unable to bite his tongue any harder without the risk of losing it, he reminded Roger, as calmly as he could, which wasn't very calm at all by that point, that the only reason Liz had been driving in the first place was that Roger, who was supposed to be visiting *them* that day, had changed his mind at the very last minute and asked Liz to make the journey herself. Before Roger could protest, Danny went on to remind him of how he had consistently refused to come and see them over the years, with a repertoire of half-arsed excuses ranging from car trouble and common colds to random bouts of unspecified fatigue, when the real reason he never wanted to visit, at least as far as Danny was concerned, was that he was ashamed: ashamed to see his daughter married to a man like Danny, ashamed to see her living in a cramped apartment in Tower Hamlets instead of some leafy suburb in Hampstead, and ashamed by just how far she'd strayed from the perfect future he'd imagined for her.

The two of them didn't speak to each other for the next six months, although Carol called every now and again for an awkward chat about the weather and to talk to Will (always a one-way conversation). Each time she called she gave a different excuse as to why Roger couldn't come to the phone, as if she still felt the need to pretend that everything was fine despite being present when the two men were publicly expressing their dislike for each other while blocking everybody's access to the canapés.

Then a few months ago, Danny had received a text message from Carol asking if he and Will would meet her and Roger for a chat. He'd told her they were more than welcome to come to the flat, but she politely deflected the invitation and suggested they meet at a Pret A Manger near Old Street Tube station instead, presumably because it was neutral territory.

Danny had no idea what they wanted to talk about. He wasn't about to apologize to Roger and he knew that Roger felt the same way. A part of him feared they'd try to convince him that Will would be better off with them, which did little to put him at ease as he hugged Carol, nodded at Roger, and watched Will embrace them both, but his suspicions couldn't have been further from the truth.

Roger had family in Melbourne, something he was quick to remind people of whenever the sky was anything but blue, and he and Carol had been talking about moving to Australia for as long as Danny had known them. He never thought they'd actually go—neither did Liz, for that matter—but Carol had called the meeting to let them know that they were doing just that. Danny wasn't particularly saddened by the news—their presence in his and Will's life had always been minimal—but the announcement reminded him of just how alone he was. His wife was gone, his father was gone, his mother was as good as gone, and now his wife's parents were going. All he had left was Will, and it often felt like he was gone too. He hadn't spoken a single word since emerging from a coma after three of the most agonizing days of Danny's life, and nobody knew why. Pediatricians, psychologists, psychiatrists, and speech-language pathologists all had different opinions about his condition. Some thought the head injury he'd sustained in the crash had permanently impaired his ability to form coherent sentences. Others suggested that while his capacity for speech remained intact, Will actively chose not to talk due to the trauma he'd experienced as a result of the accident and the subsequent loss of his mother. Nobody could agree on the cause, but everybody seemed to have an opinion on the matter, much of which came from outside the medical sphere. Reg, his landlord, believed a hefty backhand would get him talking; one of the plasterers at work swore by the powers of hypnotherapy; and once, while Danny was rummaging through the reduced bin at the supermarket, a woman with a long gray ponytail whom he'd never met before suddenly appeared beside him and casually asked whether Will was taking enough ginkgo biloba. Whatever the solution, if there even *was* a solution, Danny had long since given up any hope of finding it. He'd given up hoping for anything except the day when he would wake up and feel something other than a desire to close his eyes again and never reopen them.

He picked up the frame that sat on the coffee table beside the couch. Inside was a photograph of Liz that Danny had taken on a summer's day in Hyde Park a couple of years ago. She was lying on the grass with her cheek on her arm and a smile on her face that was sleepy from the warm weather and the bottle of red they'd been sharing. The dress she was wearing, patterned with flowers, had barely left the wardrobe that year due to an even wetter summer than usual, and that was the last time she'd worn it before returning it to the hanger from which it still hung. Any hint of his wife had long since left the garment, but Danny still found himself checking on occasion, burying his face in the dress and breathing in whatever microscopic trace of her scent might still be trapped within the fibers of the fabric.

He ran his thumb across her cheek and smiled. Then, as he clutched his wife to his chest, Danny's shoulders began to shake as he sobbed as quietly as he could.

CHAPTER 6

Will was in the middle of not eating his breakfast when somebody banged on the door so hard that the neighbor's letterbox chattered in protest. Will looked nervously at Danny, who looked towards the hallway. Neither of them moved, Danny's arm suspended with his mug halfway to his lips and Will's fingers tightening around his slowly wilting piece of toast.

Danny didn't need to answer the door to know who was on the other side. Only one person knocked like that. Only one person knocked at all. Normal people just used the doorbell, but Reg wasn't a normal person. Whether he was even a person at all was a point of ongoing debate among those unfortunate enough to know him. Many saw him as a different species entirely, and one that should have been extinct, not just because of his poor diet, his high blood pressure, and his questionable life choices but because the world would simply be a much nicer place without him. Reg was the sort of person who would run into a burning orphanage and come out with the furniture, the sort of person who didn't cheer for the good guy or the bad guy because he wanted everyone to die, the sort of person who would puncture a football that strayed into his garden before chucking it back over the fence, and the sort of person who wouldn't think twice about puncturing tenants who didn't pay their rent on time—tenants like Danny, who hadn't paid his rent for the last two months.

The banging stopped and the radio filled the silence with an idiotic jingle for a local car dealership.

"I think he's gone," whispered Danny, his eyes moving around the room as if he were tracking a fly. The moment he took a sip of tea, however, he almost threw it over himself as the banging resumed with renewed intensity.

Worried that the door was about to leave its hinges, and worried that Reg would bill him for the damage, Danny got up and cautiously peered through the spyhole.

Reg was standing in the corridor, his sagging body held upright by a pair of grubby elbow crutches. He leaned into them with his back slightly arched and his arse jutting out, a stance that always reminded Danny of a posturing gorilla, although gorillas were generally friendlier than Reg and only attacked when threatened. His cheeks were more flushed than usual, the broken lift having forced him to take the stairs, and on his head sat an unwashed flat-cap saturated in decades of pomade. Standing behind him was a towering square-headed man with short-cropped hair as black as his suit. The two men looked comically distorted through the fish-eye lens of the spyhole, and Reg didn't look any less distorted when Danny opened the door.

"About fucking time," said Reg as he barged past Danny. "Get the kettle on."

"Reg," said Danny. He craned his neck to meet the eyes of the other man. "Dent," he said, moving out of the doorway so Mr. Dent could enter.

Will watched the men trickle into the living room, his limp piece of toast still pinched between his fingers. Mr. Dent took Reg's crutches and helped him into Danny's chair before seating himself beside Will. Danny hovered like a nervous shop assistant as his son disappeared in Mr. Dent's shadow.

"I don't hear the water boiling," said Reg.

Danny leaned into the kitchen and flicked on the kettle, his eyes still on his uninvited guests.

"What's that, then?" said Reg, pointing at Will's toast. "Peanut butter?"

Will looked down at his breakfast and then back up at Reg. He nodded.

Reg looked at Mr. Dent, who slid the plate away from Will and parked it in front of his boss.

"Not bad, that," said Reg, flecks of toast and peanut butter spattering the table as he spoke. "I prefer the smooth stuff, though, being the smooth bastard that I am."

Danny laughed, but not too much.

"Still giving it the silent treatment, is he?" said Reg, nodding at Will, who began to fidget beneath his gaze. Danny shrugged and mumbled something in the affirmative.

"You got it lucky," said Reg. "Seen and not heard, as it should be. You want to meet my youngest, right little gobshite that one, can't never shut him up. Just like his mother, drives me fucking daft."

Reg wiped his mouth with the back of his hand and wiped the back of his hand on his trousers. He took out a packet of Superkings and slipped a cigarette between his lips.

"We don't—" Danny cut himself short, but not short enough.

"We don't what, Daniel?" said Reg, leaning into the flame that Mr. Dent was holding out for him.

"Nothing."

"We. Don't. What. Daniel?" Reg's tone implied that Danny would be sorry if he made him ask a third time.

"We don't . . . I mean . . . it's just . . . it's a nonsmoking house."

"Is it?" said Reg. He took a deep drag of his cigarette and sent a cloud of blue-gray smoke towards Danny. "Remind me again, Dan, because I'm obviously fucking stupid, but whose house is this?"

"It's yours," said Danny.

"See, that's what I thought too, but the way you were talking made it sound like it was *your* house. Do you see where my confusion arose?"

"Yes, Reg. Sorry, Reg."

"So, whose house is this?"

"It's your house."

"And who makes the rules?"

"You do."

"You're fucking right I do. So don't you ever tell me what to do in my own fucking house again. You got that?"

"Yes, Reg."

"And anyways," said Reg, tapping ash into Danny's tea, "I wouldn't worry about a little passive smoking. Not if I were you. If I were *you*, I'd be more concerned about the other things in life that could pose serious risks to your health. Like, oh, I don't know, not paying your rent on time, for example."

"Will, why don't you go wait for Mo downstairs," said Danny. Will hesitated. "It's okay, mate. Me and Reg just want to have a little chat."

Will grabbed his schoolbag and shuffled out of the room.

"Sit down, for fuck's sake," said Reg, pointing to Will's recently vacated seat. "You're making me nervous."

Danny sat down at the head of the table, as far from Reg and Dent as possible.

"Look, Reg, I know why you're here and—"

"*I* know why I'm here. I'm here because this is my fucking flat. What I *don't* know is why *you're* still here."

"I appreciate you've been patient, really, but things have just got a bit on top of me lately and, well, that last rent hike was . . . it was quite steep and I wasn't really, I mean I hadn't factored in . . . I just wasn't expecting it. Not so soon after the last one."

"Yeah, well, that's inflation for you. Don't blame me, blame the economy."

"I know, but, well, with all due respect, inflation's, like, what? Three percent? And my rent went up *twenty* percent, so—"

"Admin," said Reg.

"What?"

"Admin."

"Admin?"

"Is there a fucking echo in here, Dent?" said Reg.

Mr. Dent shook his head.

"Admin," said Danny. "Right. Of course. It's just . . . we haven't got the same money coming in like before. Not since . . . you know . . ."

"Terrible thing, what happened to your Liz," said Reg. He took another drag on his cigarette. "She was a good girl, that one, I liked her a lot. But at the risk of sounding like a heartless old twat, which, come to think of it, is exactly what I am, your loss, while tragic, don't change the fact that you owe me two months' rent."

"I know, Reg, I know, and you'll get it, I promise."

"Oh, I know I'll get it. There's no question about that. The only question is *how* I'll get it. See, I prefer cash, but Dent here, he's a bit more, let's say *old-school* when it comes to recouping expenses. Ain't that right, Dent?"

Mr. Dent nodded. He opened his jacket to reveal the head of a well-used claw hammer poking out of his inside pocket. Danny's chair creaked as he squirmed in his seat.

"So what's it going to be?" said Reg.

"I'll get the money, Reg, I swear. Just give me a bit more time."

Reg took a contemplative drag on his cigarette. Nobody spoke for a moment.

"I wouldn't normally do this," said Reg finally, "but in light of your . . . extenuating circumstances, I guess it's only right that I show a little compassion. I might not be pretty, but I ain't a monster. Still, consider yourself lucky that your wife copped it. Otherwise I wouldn't be half as understanding."

Danny clamped his teeth together and forced himself to nod.

"I'll give you two more months to pay everything you owe. Otherwise you and that boy of yours will be looking for a new place to live. I have to warn you, though, Dan, finding a flat with disabled access is a lot harder than you think."

"Thanks, Reg," said Danny. "I appreciate it."

"So you should, Dan, so you should. As for interest, I'd say thirty percent is more than reasonable, wouldn't you?"

Danny winced as he struggled to swallow his objection. "More than reasonable," he said.

"Good. That's settled, then," said Reg. His cigarette hissed as he dropped it into Danny's tea.

"I don't mean any disrespect, Reg, but I've got to get to work. See, there's this new guy, he's a Russian, Vitali or something, and if I'm late—"

"Two sugars."

"What?"

"In me Rosie," said Reg. He took another bite of toast.

"Right. Your tea. Of course. Sorry."

Danny disappeared into the kitchen, cursing under his breath as he rushed to make Reg's tea.

"And don't be stingy with the milk!" yelled Reg.

Danny ran across the building site, his hard hat bobbing on his head as he ducked and dodged behind girders and excavators. He made for Ivan, who was busy unloading bags of cement from the back of a lorry.

"Danny Boy," said the Ukrainian. "Alf is looking for you."

"Did he say why?" said Danny, struggling to catch his breath.

Ivan shrugged. He was the only man Danny knew who could shrug with a bag of cement on his shoulder.

"Danny!" yelled someone from across the site. Danny turned to see a red-faced Alf gesticulating angrily from the doorway of his prefab office. "Get your arse in here now!"

Alf was sitting behind his desk, furiously clicking a retractable pen against his mouse pad.

"All right, Alf," said Danny, the floor flexing beneath his feet as he stepped into the office and took a seat opposite Alf. "What's up?"

"What did I tell you yesterday?"

Danny removed his hard hat and nervously ruffled his hair. "You told me to go and work on cement with Ivan," he said.

"Don't mug me off, Dan, you know what I'm talking about," said Alf.

Danny sighed. "All right, Alf, look, I'm sorry, really, but seriously, mate, it wasn't my fault. See, me and Will were eating breakfast when somebody starts hammering the door down, right? So I go to investigate and who's standing on my doorstep but Reg and his giant lummox of a bodyguard. Next thing I know they're sitting at my dinner table and refusing to leave until the teapot's empty."

"Go on," said Alf.

Danny frowned. "Go on what?"

"Finish your story."

"I just did."

"That's it?" said Alf. He looked like he was battling a sudden onset of brain freeze. "That's your excuse? You were late for work, *again*, even though I explicitly warned you *not* to be late again just twenty-four fucking hours ago, and the reason for that, if I'm understanding you correctly, and I really hope I'm not, is that you were busy supping Tetley with your landlord?"

"Sainsbury's Basics," said Danny.

"What?"

"It wasn't Tetley, it was Sainsbury's Basics. We can't afford Tetley."

"Unbelievable," said Alf, closing his eyes and pinching the bridge of his nose.

"They're actually not that bad once you get used to them."

"I'm not talking about your fucking tea bags!" yelled Alf, mashing the desk with his fist.

"Right. Sorry," said Danny.

Alf shook his head. "You should have listened to me, Dan. I was trying to help you."

"I did listen, but what was I supposed to do? The guy had a hammer, for Christ's sake. No, not a hammer, a *claw* hammer. Why the hell does he need the claw? What's he planning to do with the claw? I don't know, and I don't want to find out, so when a man with a claw hammer tells me to make him a cup of tea, I'm not exactly going to say no, am I? I'm sorry, Alf, really, but it's not like I had much choice in the matter."

"Yeah, well, neither do I," said Alf.

Danny's eyes narrowed. "What's that supposed to mean?"

"You know what it means."

"No, Alf, I don't."

"It means the word's come down is what it means."

"What word?"

"I've got to let you go, Dan."

"That's seven words."

"Don't make this harder than it already is," said Alf.

"Who's it come down from?" said Danny, looking around as if the culprit were hiding somewhere nearby. "That Russian wanker?"

"It don't matter who," said Alf. "The decision's been made. Clear out your locker, they want it emptied ASAP. You still got two weeks' holiday, so consider that your notice period." Alf did his best to look at everything else in the room except Danny.

"Four years I've worked for you, Alf. Four fucking years. And have I ever, *ever* let you down in all that time? Even once?"

"It's not up to me, Dan. I wish it were, but it ain't. It's this new management, mate, they're ruthless. They'd replace their own grandmothers if they could find a cheaper model. It ain't just you. Nobody's safe, not even me."

"You can't do this," said Danny. "I need this job. I *really* need this job."

Alf sighed like a man who regretted every life choice that had led him to this moment.

"I'm sorry, Dan," he said. "There's nothing I can do."

CHAPTER 7

Danny rammed his meager belongings into a grocery bag with such force that he punched a hole through the bottom of it. Cursing, he shook the bag from his arm and watched it float to the floor before kicking it and cursing again when he got his foot stuck in the hole. Yanking the tangle of plastic from his shoe and screwing it into a ball, he sat on one of the benches scattered around the locker room and buried his head in his hands. Ivan entered a few minutes later and took a seat beside him.

"Alf, he just tell me," he said.

Danny nodded but said nothing.

"You know, I have cousin," said Ivan. "He owes me favor. If you like, I call him now. He teach Alf lesson." Ivan made a gun with his fingers and pretended to shoot himself in the head. "Boom. You know?"

"It's not Alf's fault," said Danny, looking up, "but thanks for offering to murder him for me. That really means a lot."

"How about job?" he said. "You need new job? I know many people."

"Don't take this the wrong way," said Danny, "but, well, I need something . . . legitimate."

"Legitimate?" said Ivan, frowning. "What is this word *legitimate*?"

"Precisely."

Ivan nodded without really understanding.

"I'll be okay, don't worry. This isn't the only site around, there's plenty of stuff going up these days. There's that new office in Brunswick, there's that big thing going on in Farringdon, there're loads of opportunities around town. The hard part won't be finding a job. The hard part will be knowing which one to take."

"Wait," said Ivan as Danny turned to leave. He opened his locker and pulled out something shaped like a brick and wrapped in tinfoil. "Here," he said, handing the parcel to Danny. "From Ivana."

Ivan's wife had baked him a walnut cake every few weeks since Liz had died. Danny secretly looked forward to those cakes more than anything else in his life (apart from the day that Will finally started talking again, if that day ever came), not just because they tasted incredible but because they reminded him, in his darkest moments, when Will was asleep and the flat was quiet except for the sound of his own unwanted thoughts, that even though he often felt completely alone in this world, he wasn't. Not as long as there was cake in the kitchen.

"Smells amazing," said Danny, lifting the parcel to his nose. "Thanks, Ivan. And please thank Ivana for me."

"Ivan!" yelled Alf from somewhere outside.

"You better go," said Danny.

Ivan nodded but didn't move. "You will be okay?" he said.

"Of course, mate. Don't worry. I'll have a new job in no time. Just you wait."

Aside from a holiday to Margate when he was seven (a trip he only remembered because his mother had left him on a teacup ride for close to an hour while she went to the pub) and a couple of trips to Brighton, once with Liz when they were teenagers and once with Will when they were parents, Danny had never left London in the twenty-eight years since he was born. He was therefore fairly confident that he knew his hometown better than most, but over the course of the following fortnight Danny saw more of the capital than he'd ever seen before. He passed through almost every borough and he traveled through every fare zone (including zone nine) and during that time he saw countless parts of the city that he hadn't even known existed until then (including zone nine).

Not wanting to worry Will with the news of his recent sacking, and not wanting to suffer the embarrassment of having to explain it, Danny continued as normal, dressing in his tatty work scruffs every morning and making breakfast for them both before setting off for a long and fruitless day of job hunting once Will had gone to school.

He started out by visiting the larger construction outfits in central London, the ones that were shaping the city's skyline with gherkins, tins of ham, and other buildings that were inexplicably designed to resemble things you'd usually find in a kitchen cupboard. Next he tried zone two, first hoping to find work on the various skyscrapers popping up in Canary Wharf and Docklands and then, when that failed, trying his luck in Greenwich, where several new housing developments were being built. The farther out from the center he journeyed, the more the opportunities dwindled as large-scale construction projects gave way to compact starter homes and domestic renovations.

He even offered his services to an elderly man who was covered in almost as much paint as the garage he was slapping it onto, but no matter how far he traveled or who he spoke to, the story was always the same. Nobody needed what Danny could offer because Danny couldn't really offer anything. He wasn't a plasterer or a carpenter. He wasn't a roofer, or a tiler, or a bricklayer. He didn't know how to weld, and although he knew the basics of wiring and pipes, he wasn't a qualified sparky or plumber. Danny dug holes. He carried bricks. He mixed cement. He hammered nails. And he was good at all of those things. The only problem was that so were loads of other people. He had nothing to set him apart from the masses of unskilled laborers who were also looking for work, nothing that could give him even the slightest edge over anybody else in the labor market. He had no training aside from a basic one-day first aid course he'd done years ago and could no longer remember anything about, and he had no qualifications of any kind except for a school certificate in art (C-) and another one in geography (D). Over the years he'd seen countless adverts for classes and workshops and apprenticeship schemes in everything from joinery and window fitting to quantity surveying—courses that would have given him the requisite skills to improve his own career potential and prove to an employer that he was capable of more than just shoveling cement—but time and again he'd found some excuse for not putting his name down, whether it was because he was too busy, even when he wasn't, or because he didn't have the money, even when he did. He'd always known in the back of his mind that this day was coming, he just didn't know that it would come so soon. Now here he suddenly was, in debt, out of work, and in serious danger of learning the hard way just what Mr. Dent planned to do with that claw hammer unless he found a job, and fast.

He thought about applying for Universal Credit until he saw how long the waiting time was. He needed money now, not five weeks from now, so Danny began to look for work wherever it might be available. Supermarkets. Warehouses. Offices. Haulage companies. Factories. Takeaways. Newsagents. Fast food outlets. Clothes shops. Bakeries. Department stores. Cleaning companies. Waste disposal outfits. Butchers. Jewelers. Restaurants. Sandwich shops. Mobile phone shops. Pet shops. Cinemas. Bookshops. Hairdressers. Art galleries. Zoos. Cemeteries. Taxi companies. He even applied to work as a parking warden, a job that he and everybody else in the world found indefensible (including many parking wardens); but whether it was because people weren't hiring, or because his CV could have fit on a Post-it note with plenty of room to spare, Danny couldn't find a job anywhere.

Another challenge he hadn't foreseen was looking for work as a single parent. It wasn't something he'd ever thought about before because, well, he'd never needed to, but now he realized just how difficult it was. Will was never alone when Liz was alive because she and Danny would coordinate their schedules in such a way as to ensure that one of them was always free to look after him, and Danny had since adjusted his hours so that he now left the house after Will had gone to school and was home in time to make dinner every day. He needed a job with similar hours because he couldn't afford a regular babysitter (like most people in London) and he wasn't willing to leave his son alone for long periods of time (as much as Will would have liked that). Several times he saw adverts looking for night security staff with immediate starts, and many service industry jobs he found—waitstaff, bar work, call centers— required little more than a pulse and a readiness to work nights. Danny was often more qualified for these positions than for many of the others he came across, but as much as he needed a job, he had no choice but to rule them all out.

Two weeks after getting fired, Danny was wandering around Islington when he noticed a small scrap of paper taped to the inside of a murky shop window. It was a handwritten advert for a full-time assistant, although Danny's first thought when he peered through the window was that the place wasn't a shop at all but a front for something dodgy like an organ farm or a meeting place for the Flat Earth Society. Staring back at him were various crooked and sun-bleached mannequins adorned in a bizarre array of costumes, from clown masks with serrated teeth and blood-splattered surgeons' aprons to a black PVC bondage outfit complete with bright-orange ball gag. It was only when Danny stepped back and read the sign above the door that he realized he was standing outside a costume shop. Quickly smoothing his clothes with his palms and running his hands through his hair, he checked himself in the window and made his way inside.

The shop smelled like a lost property office and looked like a thrift store, albeit one that received the bulk of its donations from dominatrices, circus performers, and Burning Man attendees. It was also eerily quiet, and as Danny made his way past the various racks that led towards the counter at the back of the shop, all he could hear were the creaks of the floorboards and the murmur of the street, which suddenly felt much farther away than it actually was.

"Hello?" said Danny. He peered over the counter and into the open storeroom behind it. He waited for an answer, but not for very long because the place gave him the heebie-jeebies and all he wanted to do was leave, which was just what he was about to do when a pirate leapt up from behind the counter the moment his back was turned.

"Ahoy there!" shouted the pirate.

Danny screamed and spun around to find a man with an eye patch and a stuffed parrot on his shoulder. It wasn't a stuffed toy that resembled a parrot but an actual stuffed parrot, and a badly stuffed one at that.

"Sorry, matey," said the man in a deep and gravelly voice that seemed at odds with his youthful face. "Didn't mean to scare ya."

"Then why did you jump out like that!" yelled Danny.

The man thought about this for a moment. "Okay, fine, I did mean to scare you, but it just gets so boring in here," he said, his pirate accent giving way to a Bristolian twang. "You're the first customer we've had all day. All week, actually."

"I'm not surprised," said Danny, kneading his heart.

"Seriously, I think I would have gone mad by now if I didn't have Barry to keep me company."

"Barry?"

The man nodded at the parrot on his shoulder.

"Right," said Danny.

"What can I do for you anyway?" said the man.

"I'm looking for—"

"No, wait, let me guess. I'm usually pretty good at this. Let's see, it's too late for Easter, it's too early for Halloween, it's definitely too early for Christmas, which means that you want a costume for . . . a tarts and vicars party!"

"No, I'm—"

"Cops and robbers? It's cops and robbers, isn't it?"

"No—"

"Murder mystery?"

"Look—"

"Got it!" said the man with a click of his fingers. "You're looking for a jumpsuit for your sister's disco-themed birthday party."

"I haven't got a sister," said Danny.

"Is it a costume funeral by any chance?"

"Is that even a thing?"

"You'd be surprised."

"Look, I don't want a costume. I'm here about the advert."

"What advert?"

"In your window," said Danny. He pointed towards the front of the shop. "For the job."

"Oh, yes, the job! Sorry, I put that up so long ago that I forgot all about it."

"Well, you can finally take it down because here I am." Danny presented himself with a showroom flourish of the hands.

"I really need a woman," said the man.

"I bet," said Danny, looking the man up and down.

"No, not *me*. *I* don't have any trouble with the ladies. The boss is looking for a woman. For the shop."

"Where's the boss?" said Danny, looking around.

The pirate readjusted his eye patch. "I'm the boss," he muttered.

"Right," said Danny. He shook his head and turned to leave.

"You want to leave your number?" said the man.

"What for?" said Danny.

The man shrugged. "Maybe we can go for a pint sometime. You know, the three of us." The man nodded again at Barry.

Danny pointed to the street. "I really have to go," he said, backing out of the shop.

The man sighed as he watched the door creak shut behind Danny.

"Nice one, Barry," he said.

Barry said nothing.

Danny came home that evening to find Will on the couch watching television and a letter on the table. Fearing it might be another bill, Danny ignored it while he changed out of his "work" clothes, took a long shower, and made himself a cup of tea—three things he always did when he used to work on the building site, and three things he'd continued to do in order not to arouse suspicion—before collapsing into the armchair and peeling open the envelope, carefully, as if it might explode.

It wasn't a bill, which came as some relief, but he knew it couldn't be good news either when he saw the name of Will's school on the letterhead. Either Will was in trouble or the school wanted money for something. He quietly hoped that Will was in trouble.

"Another school trip?" he said, as much to himself as to Will. "Where to this time?"

Will continued to watch *Top Gear* while Danny continued to read.

"Stonehenge? You've already been there, with Mum, remember? It didn't cost fifty quid neither. You don't want to go again, do you?"

Will shrugged.

"I mean, it's okay if you do," said Danny. "It's totally up to you. The place hasn't changed since the last time you were there, it still looks *exactly* the same, but if you honestly think that you're going to learn something from this trip that you didn't learn last time, and I know you learned a *lot* last time, so much so that I distinctly remember thinking that this kid is literally an expert on Stonehenge now, and probably never needs to go again—if you really want to go all the way back there to look at the same old pile of rocks, then that's absolutely fine with me, you just say the word. Or not. I mean, you don't need to say anything. I just mean . . . you know what I mean."

Will didn't respond. Danny stared at the letter.

"Look," he said. "I'll tell you what. Why don't *we* go sometime, just you and me? It'll be fun, we can make a day of it, like you and Mum did. How's that sound?"

Will shrugged again while somebody blathered on about horse-power in the background.

"Great," said Danny, trying to ignore the lack of enthusiasm. "I'll go get dinner on."

Closing the kitchen door behind him, Danny scanned the letter again in the hope that a second read might reveal some previously overlooked detail that would exempt him from having to pay. Finding none, he slid the letter back into the envelope and tossed it into the bin.

The weather was bright and the park was full of retired people lounging in beach chairs, young parents pushing strollers, office workers eating their lunch or soaking up the sun, and groups of students chatting in circles on the grass.

Danny was sitting on a bench in the shade, his eyes fixed on his phone as he scrolled through endless pages of job advertisements.

"Experience required, experience required, experience required," he muttered to himself as he made his way through the list. Every job he came across, no matter how menial or self-explanatory, seemed to require some level of experience. A shop assistant required experience. A bingo hall cleaner required experience. Even a dog walker required at least two years' experience in walking progressively larger dogs "up to the level of Alaskan malamute," according to the advert (Danny guessed the starting level was probably a Chihuahua or a shih tzu or something, although the advert didn't specify).

He opened his e-mails to find two rejections waiting in his inbox and another one in his spam folder. He also had an e-mail from a woman called Svetlana who thought his Facebook profile was very attractive despite the fact that he didn't have a Facebook profile.

Danny sighed and put his phone away before slowly making his way through the park. Noticing a crowd up ahead, he saw that people had gathered to watch the same street performers that he and Will had come across a couple of weeks previously. The man with the cat on his shoulder was there, performing for a group of people who were filming him on their phones. Another sizable crowd had gathered around the magician, and the nut juggler and the chicken man were also present, along with several other performers that Danny hadn't seen before, including a mime, a one-man band, a violinist, and a human statue who was doing her best to ignore the children who were doing *their* best to piss her off.

Danny stood and watched for a while, marveling once again at their ability to earn what appeared to be good money by dressing up and making total fools of themselves. He could see why certain acts were popular, namely the ones that exhibited some kind of actual talent, such as the violinist, or the magician, but he couldn't for the life of him figure out how even the worst performers were leaving the park with more money than they'd arrived with. The only music the one-man band seemed to be making was that of the purely accidental variety, randomly flapping and kicking and twitching in the hope that at least one of his limbs struck a corresponding chord, and the man dressed up like a squirrel spent more time picking his oversize nuts off the ground than he did trying to juggle them. Danny had lost count of the number of jobs he'd come across that morning alone that he wasn't eligible for because he didn't have enough experience—yet here were people who were making not just a living but a *decent* living when they clearly had no clue what they were doing.

And just like that, Danny had an idea.

CHAPTER 8

"I bid thee welcome, weary traveler!" said the man behind the counter at the costume shop. That, at least, was what Danny assumed he said, although it was difficult to know for sure because the man was dressed in a full suit of armor and his voice was muffled by his helmet. "Aha, you again," he said, lifting his visor as Danny approached.

"Where's Barry?" said Danny, looking around.

"Currently rented out to an elderly widower called Graham."

"Never too old for a pirate party, I guess."

"He didn't hire the costume," said the man. "Just Barry."

"What—"

"I didn't ask. We need the money."

"Got it," said Danny. "What's the cheapest costume you have?"

"Allow me to cast an eye over yonder bargain rack and honor thy request, good sir," said the man. He turned to the rack behind him and began to rifle through the hangers. "How's about this one?" he said, taking a suit from the rack and draping it over the counter.

Danny frowned. "Is that ... a Nazi uniform?"

"We prefer the term *historically accurate military costume*," said the man.

"It's a historically accurate Nazi uniform."

"Well, if you want to get technical about it, then yeah."

"Has anybody ever actually rented this?" said Danny.

"Prince Harry did once, I think."

"Right. I was sort of looking for something a little less likely to get me beaten up, to be honest."

The man flicked through the rack again and selected a three-piece suit with a blue tie attached. His other hand clutched a messy blond wig.

"Well?" he said.

"Well what?" said Danny.

"What do you think?"

"What is it?"

"A Boris Johnson costume, obviously."

"I said I wanted something *less* likely to get me beaten up," said Danny.

"So . . . that's a no?"

"Yes, that's a no. Who the hell wants to look like Boris Johnson?"

"Nobody," said the man. "That's why it's cheap."

"Give me something else."

The man rummaged through the rack a third time. He found a black-and-white costume and placed it on the counter, along with a mask.

"What am I looking at?" said Danny.

The man's gauntlet clacked against the counter as he drummed his fingers in contemplation. He checked the label on the inside of the costume and shrugged as much as his suit of armor would allow.

"It's a panda," he said.

"You sure about that?"

"No, but it's what the label says."

Danny stared at the costume, still not convinced. If it really was a panda, then it was the saddest panda he'd ever seen, one that had lived an exceedingly long and disappointing life full of unfaithful partners and unreliable betting advice.

"It smells funky," said Danny, his nose twitching involuntarily.

"I won't lie to you," said the man, "some kid hired it for Freshers' Week and puked all over it. It's totally clean, don't get me wrong, but it still has the faintest whiff of Jägermeister vomit."

"How much do you want for it?"

The man thought for a moment.

"Tenner?"

"I'll give you five."

"Give me a tenner and I'll throw in Boris for free."

Danny pulled a crumpled five-pound note from his pocket and slapped it onto the counter.

"I'll give you five," he said.

"Deal," said the man.

Back at the park, Danny locked himself inside one of the public bathroom stalls and commenced his graceless metamorphosis, almost putting his foot in the toilet as he struggled to remove his clothes in the cramped confines of his makeshift dressing room, and then almost doing the same with the other foot when he tried to feed his leg into the costume.

"All right?" he said to the man at the urinal who gaped at him as he emerged from the cubicle. The man nodded, unaware that he was missing the urinal by a good few inches as he watched the giant panda check itself in the mirror.

Danny wandered around the park in search of a suitable place to perform. He thought it was a good idea to keep his distance from the other acts for the time being, partly because he didn't yet have the confidence to approach them and partly because he didn't want to risk a potential territorial dispute on his very first day. By turning up uninvited without first introducing himself or asking for acceptance, he might be breaching some unknown code of honor which might or might not result in a gangland-style punishment beating or, at the very least, some unpleasant staring.

Deciding on a spot that was far from the other performers yet still close enough to keep an eye on them should he need to make a hasty getaway, Danny dumped his bag of clothes on the grass behind him and placed his open lunchbox by his feet. Lining the container with a handful of pocket change, he nervously adjusted his costume and thought about what to do next.

As if smelling his fear, a little girl suddenly appeared in front of him while her mother hovered nearby. The girl wore a yellow dress, blue glasses, and pigtails in her hair, but despite her seemingly innocent appearance Danny couldn't help but feel the slightest bit intimidated as she silently waited for him to do something. Unable to juggle or play the guitar, and without a cat to balance on his shoulder, Danny did the only thing he could think of. He waved.

The girl continued to stare at him, her eyes wide because her glasses were thick and not because she seemed even remotely wonderstruck by the weird-smelling panda in front of her.

Already out of ideas and feeling increasingly awkward, Danny waved again. The girl looked at her mother, who smiled at Danny almost apologetically before removing her purse from her handbag and giving her daughter some money.

"For me?" said Danny as the girl approached with a pound coin pinched between her fingers. Instead of giving him the money, however, the little girl, hypnotized by the sight of the other coins in the lunchbox, quickly grabbed a handful of change and stuffed it into her pocket while her mum, still smiling, looked on in total oblivion.

"Hey!" said Danny, instinctively grabbing her arm.

The girl screamed so loudly that several people stopped to see what was happening.

"Tamara!" shrieked the mother. "Get away from her, you pervert!"

"She stole my money!" said Danny as the woman ran over and scooped her daughter up in her arms.

"The bad man touched me!" wailed the girl.

"I didn't touch her!" said Danny, addressing the bystanders, one of whom had started to film the encounter on his phone. "I mean, yes, I touched her, but I didn't 'touch' her," he added, making inverted commas, which only made him look worse.

"You're lucky I don't call the police!" said the mother.

"You're lucky *I* don't call the police!" said Danny, prodding his furry chest. "I'm the victim here!"

"Victim!" said the mother. She pointed to her daughter. "She's five years old!"

"So was the kid in *The Omen*!"

"Are you calling my daughter the Antichrist?" She looked at the man who was filming. "Did you get that? He just called my daughter the Antichrist."

"Can you please stop filming?" said Danny.

"No way," said the man. "This shit's going on YouTube."

"What's an Antichrist?" said the girl.

"Nothing, darling," said the mother. "Come on, let's move away from the bad man."

The woman marched off with her daughter in tow. The girl looked over her shoulder and paused her theatrical sobbing just long enough to flash Danny the smuggest of grins.

He picked up his lunchbox and sighed. Over half his money was gone. Before he could calculate precisely how much was missing, a little boy came out of nowhere and kicked Danny hard in the shin, causing him to drop the lunchbox and send coins flying everywhere.

Danny clutched his leg in pain and lolloped after the scattered change. The boy giggled and kicked him again.

"Stop that!" said Danny. He waved his arms at the big man in the small suit who was barking aggressively into his phone nearby, but the man was too busy telling somebody called Dave what an incompetent wanker he was to notice what his son was up to.

The boy picked up a two-pound coin and taunted Danny with it.

"Give that back!" said Danny. The boy shook his head.

"Give. It. Back," he repeated in his best Dad Voice, and this time the boy relented. He presented the coin in his little doughy palm, but when Danny tried to retrieve it, the boy jerked his hand away and hoofed him in the shin again before laughing like a maniac and running off to show his dad the two-pound coin he'd just "found" on the path.

Danny got down on his hands and knees and wearily combed the ground. He didn't see the other children approach until their shoes appeared in front of him.

"What are you?" said the owner of the red shoes, a little girl no older than six who clutched a floppy-eared rabbit to her chest.

"He's a badger, stupid," said her brother, who had the same ginger hair and freckles as his sister.

"I don't like badgers," said the girl.

"I'm a panda, actually," said Danny, climbing to his feet and patting himself down.

"I don't like pandas," said the girl.

"Can you do kung fu?" said the boy.

"Pandas can't do kung fu," said Danny.

"Kung Fu Panda can," said the boy.

"Kung Fu Panda isn't a real panda," said Danny.

"Neither are you," said the boy.

Danny had no response to that.

"Do kung fu!" said the boy.

"Yeah, do kung fu!" squealed the girl.

"No."

"Why?" said the girl.

"Because he can't," said the boy.

"Precisely," said Danny.

"You're the worst panda ever," said the girl.

"Okay. Fine. Here," he said, throwing out a pitiful combo of clumsy karate chops. "Happy now?"

"That was rubbish," said the boy.

"That was rubbish!" repeated his sister.

The boy pointed across the park. "The man over there does magic," he said.

"Good for him," said Danny.

"Make me disappear!" shouted the girl.

"I wish I could," said Danny.

"Can you juggle?" said the boy. "Another man was juggling."

"Yeah, juggle!"

"Look," said Danny, taking a coin and holding it up. "Here's fifty pence. You can have it if you promise to go away."

"Fifty pence each?" said the boy.

"I want a pound!" said the girl.

"If she gets a pound, then I want a pound," said her brother.

Danny sighed and shook the box as if he were panning for gold.

"Here," he said. "Two pounds. One for you and one for you. Now, please. Go. Away."

The children snatched the money and ran off arguing about whose coin was bigger.

Danny sat down on a nearby bench and buried his face in his hands. He wasn't sure how long he'd been there when he heard the bench creak beneath the weight of another person. He looked up and saw the young street musician rolling a cigarette beside him. His cat was perched on his shoulder wearing a stylish violet cardigan while the man wore a tatty tweed jacket, a wilting bow tie, a pink pair of corduroys, and a low-rise top hat with a pigeon feather sticking out of it. Danny thought he looked a bit like a scarecrow, but one with rolling tobacco sprouting from his pockets instead of straw.

"How'd you get him to stay up there like that?" asked Danny, nodding at the cat.

"Milton?" said the man without looking up from his rolling papers. "He climbs up there himself. He pretends he likes the view, but I know he really just likes the sense of superiority." He slipped his cigarette between his lips and extended his hand. "Tim," he said.

"Danny."

"First day?" said Tim as they shook hands.

"Last day more like."

"That bad?"

"Well, let's see, I've been called a pervert, I've been kicked in the shins, I have less money than when I started, and I only started twenty minutes ago."

"Sounds a lot like my first day," said Tim.

"Really?"

"Yeah. Well, nobody called me a pervert. They called me plenty of other things though. Tramp. Student. Wankpuffin, whatever that means." He licked his finger and dabbed his cigarette to stop it from burning unevenly. "Wait, why did they call you a pervert?"

"I touched a little girl," said Danny. He shrugged. "No big deal."

Tim took a long drag on his cigarette. "Right," he said.

"Not like *that*," said Danny. "She stole my money so I grabbed her arm and, well, it all got blown out of proportion."

"You want to be careful doing stuff like that. You could end up losing your license."

"License?"

"Your street performer's license," said Tim. Danny frowned. "You *do* have a license, don't you?"

"Obviously," said Danny.

"You don't, do you?"

"No."

"Then you better get one before the coppers come sniffing. They love to catch illegal street performers, they think we're all just glorified beggars or something."

"Aren't we?"

"Look, just get yourself a license. Without it, you're nothing but a weirdo in a costume."

"And with it?" said Danny.

Tim shrugged. "You're a weirdo in a costume with a license."

"How long does it take?"

"Five or six weeks, give or take."

"Five or six weeks!" said Danny.

"Maybe four if you're lucky."

"I can't wait that long."

"You don't really have a choice," said Tim. "Not unless you know someone who can get you a fake one." He blew a smoke ring that was fatter than a doughnut. Milton looked like he was contemplating eating it. "What's the rush anyway?"

"I'm two months behind on my rent and if I don't pay up in the next six weeks, then my landlord, who I'm starting to think is either Satan himself or a very close relative, is going to beat me up and evict us."

"Us?"

"Me and my son, Will."

"And you thought that becoming a . . ." Tim pointed at Danny's costume but failed to find the right words.

"Panda bear."

"Right," said Tim, not entirely convinced. "You thought that becoming a panda would somehow solve this problem?"

"No, I thought that doing overtime at the building site I worked on would somehow solve this problem, and then when I got fired

I thought that finding another job would somehow solve this problem, and then when I couldn't find another job I saw how much you guys were making and thought fuck it, what do I have to lose?"

"Some of these guys make a pretty decent living, that's true, but they're talented and they work hard. You're going to need a really good act if you want to survive in this business."

"Act?"

"Yeah, like, what do you do?"

"This isn't enough?" said Danny, gesturing to his costume.

"Yeah, it's enough to get you evicted." Tim took a drag on his cigarette. "Do you know how to play anything?"

"Badminton?" said Danny.

"I meant an instrument."

"Oh. Then no."

"Do you know how to dance?"

"About as well as I know kung fu."

"You can do kung fu?"

"Nope."

"Get yourself a cat, then," said Tim. "This fella right here? Money magnet. Everybody loves him. Well, except El Magnifico. He doesn't like him much."

"El Magnifico?" said Danny.

Tim pointed at the magician across the park. "That guy. Proper weirdo. Thinks he's a real wizard, like Gandalf or some shit. He tried to set Milton on fire last week."

"With what?" said Danny.

"With his mind," said Tim, tapping his temple. "Keep well clear if I were you. The guy's got more issues than *Reader's Digest*."

"Thanks for the warning," said Danny.

"No problem. And get your act together. Literally."

Tim flicked his cigarette away. "Oh, one last thing," he said, standing up. "Always keep an eye on your stuff. People nick anything that's not nailed down around here."

"Got it," said Danny. "Thanks again."

He watched the man leave, still marveling at the plump cat perched so calmly on his shoulder, until he suddenly remembered his unattended clothes. Jumping up, he ran over to where he'd left them, but it was too late. His bag was nowhere to be seen.

CHAPTER 9

Danny had once seen a man in his sixties pogo-stick the entire length of Regent Street dressed in nothing but a baggy pair of underpants that rode precariously lower the closer he got to Piccadilly Circus. Thousands of other people saw him too, but nobody paid much attention. That was one of the many things that Danny loved about Londoners. Nothing fazed them, no matter how strange, and the weirder something was, the less likely they were to give a shit about it. Or so he'd always believed. But as the doors closed behind him and the bus lurched into motion, Danny realized that his perception of Londoners might not have been entirely accurate.

He tried to act normal, or as normal as he could while riding public transport dressed as a panda, but the other passengers made it difficult to maintain that illusion, especially the teenagers who were filming him on their phones and the old lady in the oversize duffel coat who glared at him as he sat down. He thought about removing the mask at least, partly because he was sweating but mainly because he felt ridiculous, but fearing that somebody from the building site (or, worse, Will's school) might recognize him, he grudgingly left it on and tried to ignore the looks he was getting.

The bus groaned to a halt and a young, dark-haired woman got on. She was tall and slender enough to make her way through the crowded bus without disrupting anybody, but that didn't stop her from disrupting people anyway, shunting passengers out of her way as she strode down the aisle, even those who weren't in her way to begin with. Her colossal hoop earrings were larger than the bus's grab handles and she chewed her gum so loudly that the sound hit Danny long before she whacked him in the head with her handbag.

"Watch out," said Danny.

"What?" said the woman as she took the seat opposite, her already mini miniskirt riding even higher on her thighs as she did so.

"You just hit me in the head with your bag."

"And? You just hit my bag with your head, but you don't hear me bitchin' about it."

"What you got in there anyway, a brick?" said Danny, rubbing his head.

"Brass knuckles," she said. "Wanna see?"

He glanced at her fingers. They were covered in so many rings that he doubted she needed knuckle dusters. Her nails were painted neon pink and she clutched a mobile phone embossed with shiny studs spelling *Krystal.*

Danny shook his head and looked out of the window. A gang of teenagers were laughing at him and giving him the wanker sign. He stared at the floor and tried to calculate how many stops he had left.

"Why you dressed like a skunk anyway?" said Krystal.

Danny said nothing and hoped she'd leave him alone. She didn't.

"Oi. Skunk. Skunky. Skunk man. Skunkerino. Skunk-a-tron. Uptown Skunk."

"I'm not a skunk," he said with a sigh.

"No?" said Krystal, sniffing the air. "You fucking smell like one."

"No, I don't," said Danny, fully aware that he did.

"Yeah, you do. You smell like a sock full of yesterday's puke."

"Skunks don't smell like socks full of puke," said Danny, remembering that Mo had once told him that skunk spray actually smelled like an ungodly combination of burning tires and moldy onions. He briefly considered telling Krystal this but decided against it.

"Well, this skunk does," she said, pointing at Danny.

"I told you already, I'm not a skunk."

"What are you, then? A ferret with scabies?"

"No."

"A rat with Ebola?"

"Wrong again."

Krystal turned to the old lady in the duffel coat beside her. "Any ideas?" she said.

"A pervert," said the old lady, scowling at Danny.

"I think you might be right," said Krystal.

"I'm a panda. Okay? That's what I am. A panda. Got that? Great."

A burst of laughter shot from Krystal's mouth so forcefully that her gum flew out and stuck to Danny like a little gray belly button.

"A panda!" she said. "Fuck me, that's a good one."

"Seriously!" said Danny, staring at the masticated wad of gum that was now firmly attached to his fur.

"Stay still," said Krystal, aiming her phone at Danny and struggling to keep it steady amid fits of laughter. She snapped a photograph, looked at it, and cracked up again.

"I really don't see why this is so funny."

"Shush," she said, muttering to herself as her fingers danced rapidly across her screen. "Look at this tragic bastard lol no wonder pandas are extinct lol who would shag that lol hashtag sad fucker hashtag pervert."

"Can you at least give me a tissue or something?" said Danny as the bus crawled to a halt and Krystal stood to leave.

"Here," she said, pulling a handful of napkins from her purse and throwing them at Danny. "Later, gummy bear. Get it? Gummy bear?"

"Hilarious," muttered Danny, plucking at his fur with a napkin as Krystal got off the bus. He looked at the old lady, who was still scowling at him.

"I'm not a pervert," he said.

Ivana opened the door, screamed, and slammed it shut again.

Danny stood in the corridor for a moment as he tried to figure out what had just happened. Remembering that he still had the panda mask on, he was about to remove it when the door flew open a second time and Ivan came barging out. Before Danny could speak, Ivan grabbed him by the throat and backed him up against the wall while Ivana repeatedly clobbered him with a broom.

"Stop," he croaked, struggling to prise Ivan's fingers off. "It's me . . . Danny . . ."

"Danny?" said Ivan, his grip relaxing slightly.

"Danny?" said Ivana. She instantly dropped the broom. "Why you dress like rat?"

"Yeah, why you dress like rat? Ivana, she hate rats."

"I'm not a rat," said Danny, removing the mask and gently massaging his half-crushed larynx. "I'm a panda."

"I almost black your eyes like panda," said Ivan, waving his knuckles in front of Danny's face. "Come, before neighbors see."

Ivan ushered Danny into the flat, which was absurdly small for a man of Ivan's size, although everything seemed absurdly small for a man of Ivan's size, even things that weren't necessarily small at all, like vending machines, and fridge freezers, and certain brands of car.

Danny nestled himself amid a sea of hand-embroidered cushions that covered almost every part of the couch. Ivan collapsed into a well-worn armchair riddled with doilies and stared at Danny as if he had some terrible news to share but couldn't quite find the right words.

"So," he said, gesturing to Danny's costume. "This is what English call 'nervous breakdown.'"

"No, this is what the English call being entrepreneurial."

"This is not French word?"

"Fine, it's what the *French* call being entrepreneurial."

"So *entrepreneurial*, this is French word for 'nervous breakdown'?"

"No."

"I am confused."

"Look, I'm not having a breakdown, okay?" said Danny as he absently scratched at the dried gum on his furry belly. "This is my job now. This is how I make money."

"People pay you to dress like idiot?"

"This idiot, he save your life!" yelled Ivana from the kitchen.

"He did not save my life!" shouted Ivan before switching to a barrage of Ukrainian. He turned to Danny. "You didn't save my life."

"I didn't say I did."

"Good. Because you didn't."

"If you say so," said Danny, struggling to keep a straight face.

"So, people pay you to dress like *panda*?" He emphasized the last word so Ivana could hear.

"Not yet," said Danny. "But they will. See, I was in Veranda Park the other day, right? You know the one with all those performers? Musicians and magicians and dancers and whatever? Well, I saw how much money they were making, and seriously, they are raking it in down there, even the crap ones, so I thought sod it, I'll give it a go myself. Today was my first day."

"And how it went?"

"Well, let's see," said Danny, beginning to count on his fingers. "I got mugged by children, I had all of my clothes stolen, there might be a video of me calling a little girl the Antichrist floating around on the Internet somewhere, a woman on the bus spat her gum at me, and I got strangled by a giant Ukrainian and whacked in the head with a broom. You know, typical first-day stuff."

"Sorry about that."

"Sorry, Danny!" shouted Ivana from the kitchen.

"I'll forgive you if you lend me some clothes. I can't let Will see me like this."

"He doesn't know you are panda man now?" said Ivan.

Danny shook his head. "He still thinks I work at the site. I don't want him to worry."

"About your mental problems?" said Ivan, tapping his temple.

"No, about our financial problems. I can't afford to pay the rent and my landlord isn't exactly the most understanding bloke in the world."

"I tell you already, you need money, I find you job. I know many people."

"Actually, do you know anybody who can get fake IDs?"

"Of course. What you need? Driving license? Passport? Sainsbury's Nectar card?"

"I need a street performer's license."

"You need license to be panda?"

"Don't ask. Can you get me one?"

Ivan shrugged. "I make some calls," he said.

"Great, thanks, Ivan. You're a lifesaver." Ivan scowled. "Sorry. Touchy subject."

Danny sneezed and fished around for one of the napkins that Krystal had thrown at him.

"Why you have *servetka* from strip club?" said Ivan.

"What?" said Danny as he wiped his nose.

"Fanny's," said Ivan, pointing to the napkin in Danny's hand. "Is strip club. In Shoreditch."

Ivana poked her head around the kitchen door and glared at Ivan, who seemed to shrink beneath her gaze. Danny looked at the napkin in his hand. The word *Fanny's* was scrawled across it in pink looping letters.

"It's a long story," said Danny. Ivan smirked. "And, no, it's not what you think."

"What? I am not thinking anything."

"Then stop smiling at me like that."

"Like what?"

"Look, just get me some clothes, would you?"

Ivan was still smiling as he heaved himself out of his chair and disappeared down the corridor.

Danny got up and went to the kitchen where Ivana was busy chopping vegetables. "That cake was amazing," he said. "I literally had to hide it from Will so he wouldn't eat the whole thing."

Ivana put down the knife and wiped her hands on her apron.

"How is he?" she said, leaning against the kitchen counter.

Danny sighed. "I wish I knew."

"And you? How are you?"

"Do you really need to ask?" said Danny, looking down at his costume. They both laughed, but their smiles quickly faded.

Ivan returned with a pair of combat pants and an Angry Birds T-shirt.

"Here," he said, handing the clothes to Danny.

"Are these Yuri's?"

"Of course," said Ivan.

"He's twelve. I can't wear a twelve-year-old's clothes, Ivan."

"You think you will fit into my clothes?"

Danny looked at Ivan and sighed. He unzipped the costume to his waist and pulled the T-shirt over his undershirt.

"Actually," said Danny, flapping his arms to demonstrate the amount of spare fabric hanging off him, "does he have anything from when he was younger?"

*　　　*　　　*

Will arrived home a few minutes after Danny, who had only just swapped his Angry Birds T-shirt for one of his own when he heard the front door open and close.

"Hi, mate," he said, popping his head out of the bedroom as Will and Mo walked into the living room. "Oh, hi, Mo."

"Hi, Mr. Malooley. How was work?"

"Work?" said Danny, blanking for a second. Nobody had asked him that question since he'd been fired. "Oh, you mean *work*. At the building site. Where I work. Yeah, it was great, thanks, Mo. Well, not great but, you know, okay. Pretty rubbish, come to think of it. Just, like, lots of digging and carrying stuff and . . . are you staying for dinner? You're more than welcome."

"Thanks, Mr. Malooley, but Will's coming to mine for dinner if that's okay. We're just picking up some video games."

"Right," said Danny, partly relieved because he didn't have much food in the house, but also slightly disappointed that he wouldn't get to spend any time with Will that evening. They might not have much to say to one another, but having a silent meal with his son was still the highlight of his day.

Will stuffed some video games into his bag and nudged Mo towards the door.

"Bye, Mr. Malooley."

"Bye, Mo. Have fun, Will." He watched the boys disappear down the corridor, but Will didn't turn around.

Danny threw the panda costume into the washing machine and filled the drawer to the brim with detergent. He turned the dial to the monster three-hour wash-and-dry setting that Liz sometimes used to use when Will was younger and had a penchant for rolling around in things that even dogs would think twice about. When the cycle finally finished, Danny cautiously sniffed the costume. It wasn't quite as bad as before, but it still smelled as bad as he imagined a bear in the wild might smell, so he pumped it with Febreze until his finger cramped up and then looked for a place to hide it. He didn't want to hang it in his wardrobe for fear that it might infect his other clothes, so he hung it on the back of the wardrobe door, which he then opened to conceal the costume in the small space between the door and the wall near the window (which he also opened).

Weary after such a long and eventful day, Danny collapsed on the living room couch and slowly ran his hands across his face. Realizing he was being watched, he turned to look at Liz, who was smiling at him from the picture frame on the coffee table. Danny smiled back.

"What the fuck am I doing, Liz?"

CHAPTER 10

Reluctant to return to the park without a license and not yet ready to face the wrath of more angry parents and their shin-kicking gremlins, Danny spent the following day at home.

Inspired by his conversation with Tim, he sat down to make a list of all the things he could do that people might want to pay him for, but ten minutes later the page was still blank, so he decided to make a list of all the things he *couldn't* do instead.

He couldn't play an instrument, that much he was sure of. Nor did he have the time to learn one, except for perhaps the triangle, which he'd briefly played in the school band before the music teacher decided it was a little above his station and demoted him to the kazoo. Even if he had been the world's best triangler, however, the Mozart of the triangle world, the Jay-Z of the idiophone, Danny highly doubted that a panda whacking a three-sided piece of metal on a string was enough to draw the crowds, no matter how skillfully he whacked it.

Magic was another thing that Danny knew nothing about, even though his father had pulled off a vanishing act that David Copperfield would be envious of, but like all true magicians, the man had never revealed his secrets and he never returned for an encore. As for juggling, Danny couldn't even catch a cold, let alone a handful of bowling pins, tennis balls, oversize nuts, or anything else that people threw and caught for the amusement of others—but he was still better at that than he was at dancing, a word he underlined twice and accompanied with several exclamation marks.

He and Liz were similar in so many ways—they both wore their socks in bed, they both liked Marmite, they both knew all the lyrics to the *Fresh Prince of Bel-Air* theme song, they both had Piers Morgan on their list of people to invite to a poisoned dinner party—but they couldn't have been more different when it came to the dance floor. Liz could move to anything. Pop. Classical. Punk. Trance. Reggae. Country. She could even dance to post-rock, something Danny didn't even realize was possible. Movement came so naturally to her that her mother used to say that she could dance before she could walk. That's why they'd signed her up for ballet lessons at such an early age, but ballet was too restrictive for Liz. She didn't have the patience or the discipline to dance according to somebody else's rules. The more rules there were, the less fun it was, and if it wasn't fun then it wasn't dancing. It was performing, and Liz didn't care about performing, which was why she became a part-time teaching assistant at the local primary school instead of pursuing a dancing career, a decision that Danny had quietly respected despite knowing what a rare and remarkable talent she was wasting.

He knew she had a gift the very first time they'd met, when Katie, a mutual friend of theirs who ended up marrying a remarkably unattractive and overweight man who, irrespective of his shortcomings, still managed to find at least three people to have affairs with, invited Liz to Danny and Katie's school disco, an event that lived long in the memories of those who attended for no other reason than that somebody actually danced that night. Kids didn't go to the disco to dance. They went to cop a feel, or to try to cop a feel, or pretend to cop a feel so they could tell their mates about it.

The dance floor was treated like a weird uncle at a birthday party: it had to be there, but everybody went out of their way to avoid it. Everybody but Liz, that is. While the other kids were fumbling in the dark or pretending to be drunk on the one sip of the one Bacardi Breezer that somebody had managed to smuggle in, Liz was busy tearing up the dance floor, much to the joy of the otherwise redundant DJ. When the music finally stopped and the teachers sent the kids home so they could get their own party underway, Katie, Liz, Danny, and his friend Mike went back to Katie's house where, in the absence of her parents, who had gone away for the weekend, the four of them drank a crusty bottle of ouzo they found in the back of the liquor cabinet, a decision that resulted in Katie waking up facedown in the flower bed, Mike waking up with two of his teeth in his pocket, and Danny and Liz waking up together, fully clothed, in each other's arms, with no recollection of how they got there but no immediate desire to disentangle themselves.

Whereas his late wife was something of a natural on the dance floor, Danny was closer to a natural disaster. His problem was simple. He had no rhythm. He could follow a beat and he could just about bob his head along to it, but everything seemed to fall apart the moment his limbs got wind of the party. His arms and legs would run riot whenever he tried to dance, kicking out here and lashing out there like a deep-sea diver suffering from the bends. They didn't obey the music. They didn't even obey Danny. The only thing they answered to was the God of Shitty Dance Moves, a merciless deity who could only be appeased by the public sacrifice of Danny's dignity, which was why he never stepped foot on a dance floor unless the rest of the room was on fire.

Still, the more he thought about it, the more he realized that whether he liked it or not—and he didn't, in the slightest—dancing was his best course of action. Unlike musicians and magicians and the various other performers he'd seen, dancers didn't require any special equipment to get an act up and running. All Danny needed was a CD player, which he had, and his legs, which he also had, at least for another six weeks or so. Gently rubbing the bruise on his shin, Danny also reasoned that children would find it much harder to land a decent kick on a moving target than they would on, say, a guitarist, or a mime, or anybody brave or stupid enough to choose the life of a human statue.

He stared at the word he'd scribbled down. *Dancing.* The sight of it made him shudder, but then he thought about Mr. Dent's hammer and his shudder turned into a full-body spasm, the type that occurs when your shirt label tickles your neck and you momentarily think it's a spider.

He was still spasming when his phone rang.

"What type of panda are you?" said Ivan.

"What?" said Danny.

"Panda," said Ivan. "What type?"

"A Chinese one, I guess? I don't know. Do pandas come from anywhere else?"

"For panda license," said Ivan. "I maybe find person who can help but they ask what type of panda are you. You sing? You dance? You play *harmoshka*? What?"

Danny stared at the pad on his lap.

"Danny?"

"I dance," said Danny. "I'm a dancing panda."

CHAPTER 11

Danny was watching TV alone when Ivan called the following evening. Will had asked to sleep at Mo's (well, Mo had asked) and Danny had reluctantly agreed, but he was glad of the arrangement when Ivan said to meet him at midnight in Peckham. Ivan offered to lend him the thirty pounds that the license was going to cost, but Danny politely refused, not wanting to be in debt to any more people than he had to be, even if one of those people was a friend. He had some rapidly dwindling savings to be used in absolute emergencies, most of which had come from Liz, or, to be more precise, Liz's parents, who used to give her envelopes full of money for every birthday, Christmas, and any other occasion they could think of. It used to annoy Liz, who rightly or wrongly perceived the gifts as some kind of jab at their less-than-impressive income and refused to spend the money, but Danny was suddenly grateful for those envelopes as he plucked three notes from the small bundle that remained and stuffed them into his pocket.

The building looked as if it had once been scheduled for demolition, something that appeared to have partly gone ahead before the council had a change of heart and decided to pardon the half-mangled eyesore. Ivan was lurking in the graffiti-riddled entrance, where somebody called ChikNwings and somebody called what looked like Bumfuzzle had been waging a war of words by tagging as many surfaces as possible. Bumfuzzle appeared to be winning.

Ivan seemed nervous, which made Danny nervous, because anything that made Ivan nervous was almost certainly worth getting nervous about.

"You okay?" said Danny.

"You bring money?" said Ivan, ignoring the question.

Danny flashed him the bills. Ivan nodded.

"You bring weapon?" he said.

"Weapon?"

"You know. Bang-bang. Stab-stab," said Ivan with accompanying hand gestures.

"No, Ivan, I didn't bring a weapon. You didn't ask me to bring a weapon."

Ivan nodded and checked his watch.

"Why do I get the feeling you're not telling me something?"

"Is fine. We go."

Danny followed Ivan into the building. He found himself in a dimly lit lobby that smelled like a well-used urinal cake. Ivan pressed the button for the lift, which seemed to work, much to Danny's surprise, although the way it clacked and clunked on the way down suggested that it probably wouldn't be working for much longer.

"Maybe it's safer if we take the stairs," said Danny.

"You want to take stairs, take stairs," said Ivan as the doors rattled open. "I take lift."

Danny peered into the gloom of the stairwell. It was so dark that only the first five steps were visible, upon which lay a bundle of clothes that on closer inspection turned out to be a person, who might or might not have been breathing. Danny got into the lift.

"So how do you know this guy?" he said.

"I don't," said Ivan. "My friend, he knows him. Well, friend of sister of friend. He do business with The Shark one time, he says he get good service."

"The Shark? That's his name?"

"Not his *real* name. Is just what people call him."

"Thanks for clearing that up," said Danny. "Why do people call him The Shark?"

"Because he likes water? I don't know. Why else would he be called The Shark?"

"Because he loans people money with insanely high interest rates? Because he eats people? Because he's a merciless, dead-eyed predator?"

Ivan thought about this for a second. "Good point," he said.

The lift lurched. Danny grabbed Ivan's arm and quickly let go when he realized they weren't about to plummet to their deaths.

"So did they look genuine?" said Danny.

"Did what look genuine?"

"Whatever your friend's sister's friend bought. What was it anyway? Passport? Driving license?"

"Dynamite," said Ivan.

"Dynamite?!"

"No. Is wrong word. Not dynamite."

"Thank God for that," said Danny.

"I mean the hand bomb. You know? You pull the thing and you throw the other thing?"

"A grenade?"

"Grenade," said Ivan. "Yes. He buy the grenade. Soviet *limonka*. Very good."

"What the fuck, Ivan? I thought this guy sold fake documents!"

"He sells many things," said Ivan, smiling. "He is 'entrepreneur,' as you say."

The lift stopped and the doors opened. Danny stayed close to Ivan as they slowly walked down a long and dingy corridor towards the only door with a light beneath it. Ivan knocked six times: thrice, then twice, then once. The sound of chains being unchained and locks being unlocked came from inside the flat before the pair were greeted by a thickset man in a black leather jacket with slicked-back hair and a bushy mustache.

"We are looking for The Shark," said Ivan. The man looked him up and down and grunted. He did the same with Danny. After frisking them both with alarming thoroughness, he grunted again, stepped aside, and motioned for them to enter.

The apartment had been completely gutted of anything with purpose or value. Absent carpets exposed rotting floorboards, rusty hinges waited in empty doorways, wires protruded from naked light fittings. Even the windows had gone, their frames now home to ugly tapestries of black bin liners and damp pieces of cardboard. Only a desk and a chair remained, both of which were occupied by a man with a mouthful of shiny gold teeth and a left eye that stared off slightly to the right.

"Mr. Dancing Bear!" said The Shark in an accent that was north of Manchester but south of Newcastle. He stood and spread his arms in welcome.

Danny, being almost as nervous as he was unaccustomed to the etiquette for greeting members of the criminal underworld, thought the man wanted a hug, which was what he gave him, much to the surprise of The Shark and much to the horror of Ivan.

"So," said the man, awkwardly adjusting his jacket. He sat back down and indicated for Danny and Ivan to do the same despite there being no chairs available. "How many would you gentlemen like?"

"How many?" said Danny. He glanced at Ivan, but Ivan just shrugged. "Well, just the one, please."

"Just the one, please," said The Shark in an accent designed to impersonate Danny but which actually sounded like Dick Van Dyke impersonating a poor Dick Van Dyke impersonator. He removed a ziplock bag from his desk drawer and gently placed it in front of him. The bag was full of pinkish pills. "Voilà."

"What's this?" said Danny with a nervous laugh.

"What's this?" said The Shark, imitating Danny again. He gave his friend a get-a-load-of-this-guy look. "It's what you asked for."

Danny looked at Ivan, hoping for some kind of explanation, but Ivan looked equally confused.

The Shark opened the bag and removed a handful of pills. He popped one into his mouth and offered them to Danny. "Here," he said. "Have a try." He winked at Danny with his dodgy eye.

"I'd really rather not."

"It's good stuff, I promise."

"Thanks, but—"

"Go on."

"I—"

The Shark pulled a Taser from his jacket pocket and placed it on the desk.

"Don't mind if I do!" said Danny. He pinched a pill from The Shark's hand and winced as it crawled down his painfully dry throat. Ivan did the same.

The Shark flashed them a dull metallic grin and swallowed the remaining pills—six or seven at least—in one hungry gulp. Suddenly he twitched and clutched his chest, and for a brief moment Danny thought that the man might be having a heart attack, but before he'd finished trying to recall his first aid training (was it "Nellie the Elephant" or "Little Miss Muffet" he had to sing while doing the chest compressions? and was it fifteen compressions to two breaths, or fifteen breaths to two compressions?), and before he was able to decide whether it was even ethical to use said training to resuscitate somebody who sold drugs and occasionally grenades for a living, The Shark reached into his jacket and pulled out a trembling mobile phone. He answered the call and started chatting to a man called Rodney while Danny attempted to confer with Ivan via several furtive glances.

"What the fuck did we just take?" said Danny's glance.

"What?" said Ivan's glance.

"I'm starting to feel a bit weird. My fingers are buzzing. Are your fingers buzzing? My fingers are buzzing."

"Why you are looking at me like that?"

"Am I having a stroke? Is that what this is?"

"Seriously, stop looking at me like that."

"I really think we should get out of here," said Danny with his eyebrows.

"You look like crazy person right now."

"What do you think? Should we make a run for it?"

"Are you even in control of your eyebrows anymore?"

"He's got a Taser though. Probably not a good idea."

"My fingers are buzzing." Ivan started rubbing his fingers. "Are your fingers buzzing?"

"Would it be weird if I started dancing? Like, right now?"

"So," said The Shark, ending his call. "Where were we?"

"Money," said his friend, who Danny had forgotten was still behind him.

"Look, I think there's been some mistake," said Danny. "I'm not here to buy . . . whatever it is I just swallowed."

"Mistake?" said The Shark. His gold teeth shone but there was no warmth in his smile. "What mistake? You asked for dancing bear, this is dancing bear." He prodded the pills with his finger. "You asked for one kilo, this is one kilo." He prodded the pills again, this time more aggressively.

"No, see, *I'm* the dancing bear. Me."

"*You're* the dancing bear?"

"Yes," said Danny. He started rocking from side to side without quite knowing why.

The Shark began to drum on the desk as if he were playing a djembe. He looked at Ivan. "Is he being funny?" he said.

"He is dancing bear," said Ivan, the floorboards straining beneath him as he too started to move to a rhythm that only he could hear.

"I'm a street performer," said Danny, stamping his feet and slapping his thighs as if he were dancing the schuhplattler. "I need a license."

"To dance," said Ivan, spinning on the spot with his arms out. "He need license to dance."

"A license to dance?" said The Shark, bobbing his head as his drumming grew increasingly furious.

"Yes," said Danny, throwing out moves he didn't even know he had in him. "A license to dance."

"We don't need a license to dance!" said The Shark, pushing his chair back and leaping to his feet. "See?" He swirled his hands around and kicked at the air before appearing to hump the leg of his friend, whose resigned expression suggested this wasn't the first time he'd been forced to endure this and wouldn't be the last time either.

"Yes, we don't need license!" said Ivan, who emphasized his point by making big fish, little fish, and cardboard boxes with his hands.

"*You* don't!" said Danny, trying and failing to do the moonwalk. "But *I* do!"

"*We* don't!" said Ivan, bobbing up and down and dancing the hopak. "But *he* does!"

"Can you help?" said Danny, throwing off his jacket and wiping the sweat from his forehead.

"Yes!" said The Shark, clicking his fingers and jogging on the spot. "I can help! But first, let's dance!"

"We *are* dancing!" said Danny.

"I cannot *stop* dancing!" said Ivan with mild panic in his voice.

"How long does this usually last?" asked Danny.

"Until the music stops!" said The Shark.

"What music?" said Ivan.

"Exactly!" said The Shark.

Danny had no idea how long they stayed and danced with The Shark. Nor was he sure what time he got home, or even *how* he got home, although he did wake up with a vague recollection of dancing with Ivan down the Rotherhithe Tunnel while cars swerved to avoid them. Danny also didn't know whether The Shark would, or could, actually get him what he needed until one morning two days later when he found an envelope stuck in his letterbox. Inside, as promised, was one street performer's license. Also inside was a note. Danny read it and smiled. *Keep dancing!* it said.

CHAPTER 12

Danny kicked his lunchbox, not hard enough to capsize it, although he felt like doing just that, but enough to jostle the coins inside. He peered into the box, hoping that the jolt had brought something substantial to the surface like a two-pound coin, or a one-pound coin, or a fifty-pence piece, or literally anything silver, but all he found was a dull bronze sea of well-worn pocket change.

He tipped the money into his hand and tallied it up before jotting down the total in the notepad he used as a ledger.

"One pound twelve," he muttered, a figure that looked and sounded pathetic even without the minus sign he then added in front of it. When combined with the rest of his takings, Danny's first official week on the job had yielded a grand total of £13.46, or 34 pence per hour, roughly the same amount that a homeless person made. That wasn't speculation either but a fact imparted by an actual homeless person that very morning, not with any kind of smugness but with a genuine air of pity. Danny had shimmied, strutted, stumbled, and sweated his way through Liz's entire collection of *Now That's What I Call Music!* CDs every day for the last seven days, and the only profit he'd turned so far was the money he saved by walking the four miles home every evening. The most he'd taken on a single day was a little over seven pounds, five of which he hadn't even earned but randomly found blowing across the park, and in that time the largest crowd he'd performed in front of wasn't even a crowd at all but a speed-walking club that had briefly paused for a breather nearby before resuming their frantic wiggling.

There was no complex formula behind his failure. He knew precisely what the problem was. He simply couldn't dance. He couldn't even dance in that endearingly crap way that made people think about calling their dads more often. Nor could he dance in that so-bad-it's-funny kind of way that made teenagers want to film him and put him on YouTube, something he found oddly insulting given the mystifying things that YouTubers found hilarious. The only person who seemed to find him even slightly entertaining was Krystal, who happened to be walking through the park at the very same time that Danny was midway through butchering the Macarena.

Noticing that she was chewing gum and fearing she might be about to land another wad of Juicy Fruit on him, Danny danced his way out of spitting range while pretending he hadn't seen her, not an easy façade to maintain considering she was standing right in front of him. The more he danced, the more she laughed, and the more she laughed, the more irritated he became, until he turned off the music, folded his arms, and waited for Krystal to leave.

"Don't stop," she said. "This is the funniest thing I've seen all day."

"I guess you didn't look in the mirror this morning," said Danny, still upset about their encounter on the bus.

"Says the man dressed like a badger's hemorrhoid?"

"I told you already, I'm a panda."

"Whatever. You want me to call an ambulance or what?" She waved her phone at Danny.

"Why would I want you to call an ambulance?"

"Because—wait, you mean you weren't having a seizure just now?"

"Very funny," said Danny. "It's called dancing, if you must know."

"No, it's called making a complete bell-end of yourself in public, if you must know."

"Yeah, well, people seem to like it."

"Do they?" said Krystal. She peered at the pathetic collection of shrapnel in his lunchbox. "They got a funny way of showing it."

"I'd like to see you do better."

"I wouldn't waste my time."

"Because you can't."

"I can, actually."

"Prove it, then," said Danny, peeling off his mask and coming face-to-face with Krystal for the very first time.

"What?"

Danny threw the mask at her and unzipped the panda outfit to reveal the sweatpants and T-shirt he'd wisely chosen to wear every day since his bag of clothes went missing.

"I'll bet you twenty quid that you can't make more money than me in the next ten minutes," he said, dumping the costume at her feet.

"I wouldn't even *wear* that thing for twenty quid, never mind dance in it." She kicked the costume and backed away in case it moved.

"Okay, fifty."

"I haven't got time for this bollocks," said Krystal. She turned and marched off down the path.

"Thought as much," said Danny, grinning as he basked in a rare moment of victory. He grabbed his costume from the grass and flicked off a cigarette butt, but before he had time to put it back on, Krystal came storming back.

"Make it a hundred," she said.

Danny smiled. "You're on."

Krystal put her handbag down and edged her way into the costume with so much caution that it seemed as if she were trying to inhabit it without the material actually touching her.

She crouched beside the CD player, plucked an album from Liz's music collection, and fed the disc into the machine while muttering to herself about what an antique piece of shit it was and how the only people who still used CD players were, for reasons that Danny didn't even try to understand, lollipop men, virgins, and people who kept snakes for pets.

Conducting a series of breathing exercises akin to a free-diver prior to descent, she held her breath, pulled on the panda mask, and stabbed the play button. Danny watched smugly from the sidelines, convinced he was already a hundred pounds richer. But as Krystal started to dance, moving to the music as if she'd been listening to that one and only track since the day she'd first appeared on an ultrasound, his smile began to fade. A crowd quickly formed as everybody who passed slowed and then stopped to watch her performance until Danny had to fight to see through them. The other performers stopped what they were doing as their own crowds peeled away to watch the dancing panda at the other end of the park. Even at that distance they could hear the cheers and applause as Krystal did a backflip that ended in the splits, but nobody cheered and clapped louder than Danny as he watched his lunchbox filling with money.

The music stopped and Krystal took a bow before ripping off the mask and throwing it at Danny.

"You were incredible!" he said as the crowd began to trickle away.

"I know. Pay up."

"How do you dance like that?"

"I'm a dancer. Pay up."

"Can you teach me?"

"I ain't teaching you nothing." She wriggled out of the panda costume and kicked it at Danny.

"Please. I just lost my job and I *really*—"

"Boo-hoo. Pay up."

"My landlord is literally going to kill me if—"

"Good. Pay up first." She jabbed her palm at Danny.

"I can't," he said, grabbing the costume off the ground and shrugging back into it.

"What?"

"I don't have it."

"We had a deal!"

"I know," said Danny, zipping up the costume, "and I'm sorry, really, but if you teach me to—"

"Well, I'm taking this, then," said Krystal, tipping the contents of Danny's lunchbox into her purse. "Twat."

She turned to leave and walked straight into El Magnifico. Standing on either side of him were the nut juggler and the breakdancer, who were trying to look as intimidating as people dressed like oversize squirrels and chickens could look.

As per Tim's instructions, Danny had kept well clear of the magician until then, but now that he was up close he could see that the man was around his age but with less stubble, more hair, and distinctly more eyeliner, not just around his eyes but also across his upper lip where a pencil mustache had been drawn. His face appeared to be modeled on that of a French waiter whose wine-pairing recommendations had just been ignored.

"Well, well, well," said El Magnifico. "If it isn't Christina."

"Well, well, well," said Krystal. "If it isn't El Magnificunt."

"I told you not to call me that."

"And I told you to stick your dick in the toaster," said Krystal.

El Magnifico sighed. "I see you haven't changed."

"And I see you're wearing my bathrobe!" She pointed at the magician's getup. "I fucking knew you had it!"

"It's not yours," said El Magnifico. "And it's not a bathrobe. It's a magician's gown."

"No, it's a bathrobe—correction, it's a *woman's* bathrobe—and it's mine."

"I'm not going through this again," he said.

"Fucker stole my bathrobe," said Krystal to Danny.

"I didn't steal anything, Christina. I told you already, my mum bought it for me."

"Why would your mum buy you a woman's bathrobe?"

"It's not a fucking woman's bathrobe! It's a fucking magician's gown! This is how fucking magicians dress!"

"David Blaine doesn't," said Krystal.

"He's an illusionist!" said El Magnifico. "Totally different dress code!"

"What about Uri Geller?"

"Do I look like a spoon bender?" he said. "Actually, don't answer that."

"Paul Daniels didn't dress like that," said Danny.

"I'm sorry, who are you exactly?" said the magician with a frown. "Oh yes, that's right. You're the new fish. I heard you fell victim to a spot of thievery the other day. Terribly sorry to hear that. You can't trust anybody these days, can you?" He let out a guilty chuckle. The squirrel and the chicken joined in. Danny's eyes narrowed. "Anyway, must dash. Always a pleasure, Christina. And you, ferret person—"

"I'm a panda."

"Then best of luck, panda man. Having seen how you dance, I'd say you're going to need it. Shazam!" He lobbed a smoke bomb at the ground and scurried off behind what he believed to be an impenetrable cloud, seemingly unaware that he was still visible.

"Such a wanker," said Krystal, her eyes fixed on the magician as the squirrel and the chicken ran after him.

"You and him were . . . ?"

"Shut it."

"Sorry," said Danny as he tried to stifle a smile. "Look, I think we got off on the wrong foot."

"That's all you got, mate," said Krystal, setting off across the park in the opposite direction to El Magnifico. "Two wrong feet."

"So teach me how to dance!" shouted Danny.

Krystal kept walking. "Lesson one," she said, spinning around and showing him the finger. "Learn to swivel."

Danny sighed and stared at the mask in his hands.

"Looks like it's just you and me," he said, pulling it over his head and slotting a new disc into the CD player.

He was just about to press play when he heard a noise coming from the corner of the park where a cluster of trees formed a small wooded area. He looked over and saw three teenagers in school uniform. They were laughing and shouting at a smaller boy with blond hair who was walking a few paces ahead of them with his head down.

Danny didn't recognize their voices, but he recognized the silence immediately.

"Where you going, Willy?" said Mark as Tony bounced an acorn off the back of Will's head. "Oi! Willy! Willy Wanka! Looking for your boyfriend?"

Will picked up his pace as the three boys lingered on his heels.

"Oi, loser!" shouted Gavin. "He's talking to you."

"You know, I'm pretty sure it's illegal to have an exposed Willy in public," said Mark.

"I think you're right, mate," said Tony.

"Disgusting behavior, that," said Gavin.

"Reckon we need to cover that thing up before somebody sees it," said Mark.

Gavin snatched Will's bag from his shoulder while Tony grabbed the hem of his coat and pulled it over his head.

"Much better," said Mark.

Will squirmed beneath the coat that the others held over his face.

"Much," said Tony.

"Still, I think we need to teach him a lesson," said Mark. "Can't have this sort of thing happening again, can we?"

"Yeah, school him, Mark," said Gavin.

Mark drove his fist into Will's stomach. Will wheezed and folded from the force of the punch.

"Again!" said Tony.

"You want another one?" said Mark, leaning close and whispering to Will through his coat. The fabric rustled as Will shook his head.

"I can't hear you," said Mark.

"I think he likes it," said Gavin.

"Tell me to stop and I'll stop," said Mark. "Just say the word and I'll leave you alone."

Will said nothing, his body limp between Tony and Gavin, who held his arms to stop him from falling, or escaping.

"Go on," said Mark. "I'll give you three seconds. One."

Will struggled, but the other boys held him firmly.

"Two. Just say stop and I'll stop."

Tony and Gavin tightened their grip as Will began to buck like a deer in a fence.

"Three," said Mark. He drew back his fist, but before he could throw it, Danny came bursting through the bushes behind them. He charged at the boys, growling and flailing his arms around as if they might just believe that a real if not uncharacteristically energetic panda was on the loose. Believing themselves to be under attack by a furry creep in a mask, however, the boys promptly scarpered through the trees, Mark at the head of the pack and his goons close behind, while Will was left to slowly untangle himself.

"Thanks," he said, emerging from his coat and straightening his uniform.

Danny nodded, his eyes fixed on the boys as he fought the urge to chase them down and shake his fist at them in a mildly threatening but ultimately pointless manner. It took a few seconds for the full magnitude of what had just happened to sink in before it finally dawned on him what he'd heard, or thought he'd heard, or hoped yet couldn't believe he'd heard. He listened hard to the silence that followed, trying to salvage some scrap of Will's voice, but all trace had gone except for whatever remained in his memory, which caused him to wonder if Will had actually spoken or whether he'd simply imagined it.

Turning to look at his son, Danny opened his mouth to say something, *anything* that might keep the conversation alive if only for a few more seconds, but even if he could find his voice, which he couldn't, and even if he knew what to say, which he didn't, Will had already gone.

CHAPTER 13

The Cross-Eyed Goat was the sort of pub where the music used to stop when a stranger walked in, until someone destroyed the jukebox with someone else's head. It was the sort of pub with telephone numbers on the bathroom walls that people actually answered if you called them. Danny knew this because he'd dialed one shortly after Liz's death, not to enquire about any of the grotty services that were scrawled in lipstick beside the number but because he was hideously drunk and just wanted somebody to talk to, even if that person turned out to be a twenty-year-old woman called Bernadette who sounded suspiciously like a forty-year-old man called Ian who worked part-time at the pizza shop down the road. The pub, which smelled like a pile of unwashed tea towels, was a second home to the various junkies, hooligans, alcoholics, and fugitives who could often be found propping up the bar or lurking in whatever corners had clear lines of sight to the door, and for some it was even a first home, like the girl who sometimes slept behind the fruit machine and kept her toothbrush in the bathroom. Still, despite its shortcomings, of which there were many, the Cross-Eyed Goat was cheap, it was close to home, and it did a surprisingly good curry on Thursdays.

Danny sat in the corner and quietly sipped his pint while a regular known affectionately as Dodgy Ken (and unaffectionately as Gropey Gary) yelled obscenities at the television, unaware that the horse race he was watching—a race he'd been persuaded to bet a large sum of money on by another regular known mysteriously as The Spatula—was in fact a rerun of the 1998 Grand National, a race in which the horse he was backing never even crossed the finish line.

Everyone turned towards the door as Ivan entered the pub. A few of the punters sized him up before wisely returning to their drinks.

"Ivana, she hurt me if I miss the *British Bake Off* program, so tell me urgent thing already," he said, pulling out a stool and taking a seat opposite Danny.

"It's Will," said Danny.

Ivan's stool creaked as he straightened up. "He is okay?"

"He's more than okay, Ivan. He spoke."

"He speak?" said Ivan. Danny didn't see his friend smile often, but now was one of those times.

"I know. I can't believe it."

Danny flinched as Ivan slapped his back so hard that he cured a trapped nerve in his left shoulder and simultaneously trapped another in his right.

"This is great news!" said Ivan. "What did he say?"

"He said, 'Thanks,'" said Danny, massaging his shoulder.

"And you say what?"

"Nothing."

Ivan frowned. "Your boy, he speak for first time in forever, and you say nothing?"

"I know, I know. I wanted to, believe me, but, well, I couldn't."

"Why you couldn't?"

"I was too shocked, I guess. I wasn't expecting it. And anyway, I was wearing the costume."

"The rat?"

"The panda."

"I thought you not want Will to know you are rat now? Panda. Whatever."

"I don't," said Danny. "That's why I couldn't say anything. I was in the park and I saw Will getting beaten up by some older kids so I ran over to help. He said thanks, I said nothing, and he hasn't spoken a word since. I don't know what to do. What if he doesn't speak again? What if that was it?"

"Is easy," said Ivan. He pulled out a foil-wrapped parcel from the plastic bag he was carrying and gave it to Danny. The Cross-Eyed Goat was the only pub Danny knew of where foil-wrapped parcels were frequently exchanged without anybody batting an eyelid. "Tell him he cannot eat cake until he speak again. If he does not speak, you eat all the cake. Is win-win for you."

Danny smiled. "Thanks, Ivan."

Offensively loud Ukrainian pop music began to blast from Ivan's pocket. He fished out his phone, looked at the screen, and cursed before answering. Ivana yelled at him for a solid sixty seconds, her voice so loud that it woke one of the locals, who lifted his head from the table and looked around in a daze with a beer mat stuck to his cheek.

"I have to go," he said when Ivana hung up on him midsentence. "Do not worry about Will. He talk once, he talk again. You will see."

"I hope you're right," said Danny. "Enjoy the *Bake Off*."

"I rather bake my *yaytsya*," muttered Ivan as he ducked out into the night.

The door had barely stopped swinging when in walked Mr. Dent. Reg hobbled in behind him as the sound of scraping chairs and nervous whispers filled the room. Everybody looked scared, even the people who made a living scaring other people. Danny eyeballed the fire exit, wondering if he could make it in time.

"The usual, is it, Reg?" shouted Charlie, the landlord.

"And another for young Daniel here," said Reg as he slowly made his way over to Danny's table. He handed his crutches to Dent and carefully lowered himself onto the stool that Ivan had recently occupied. Dent remained standing, looming over them like an over-zealous chaperone.

"That's very kind of you, Reg, really, but I was actually just—"

Before he could finish his sentence, Charlie arrived with a pint for Danny and a flamboyant cocktail for Reg that contained, among other things, a miniature paper umbrella, a colorful curly straw, a cherry in the middle, and a big wedge of pineapple on the rim of the glass. It didn't look like a drink so much as a cheap package holiday to Cancún.

Reg picked up his cocktail and clinked it against Danny's glass.

"I do love a good piña colada," he said, his lumpy tongue slithering around his lips after slurping on his curly straw. "Not a lot of people know this, but the secret to a good piña colada's in the coconut, ain't it, Dent?"

Mr. Dent nodded, being an obvious expert in such matters.

"See, most people use coconut milk, but a real piña colada's made with something called Coco López. It comes from Puerto Rico, not easy to find around here, but Charlie imports it special. He's good like that."

Reg took another sip while Danny tried to figure out exactly where this conversation was heading.

"Reminds me of my younger days," said Reg with a groan of nostalgia. "Sitting in the sun by the sea, watching the girls go by."

"In Puerto Rico?" said Danny, surprised that Reg had been farther than Slough, never mind San Juan.

"Brighton, Dan, keep up."

"Brighton. Right."

"It was a bit like Puerto Rico back then, though, depending on who you got mixed up with."

Reg plucked the pineapple from the rim and noisily sucked the flesh from the skin.

"I ever tell you how I ended up on crutches, Daniel?" he said.

"No, Reg," said Danny, his hands sliding over his knees as if he feared a physical demonstration.

"It was a bouncy castle that did it."

Danny nodded. Then he frowned.

"You fell off a bouncy castle?" he said.

"No, Danny, you pillock, I did not fall off a bouncy castle. I *owned* a bouncy castle."

"Yes. Sorry. I thought—"

"The Boogie Bounce, it was called. Shit name, I know, but that was the seventies for you. Everything was *groovy* this and *boogie* that. Anyway, there was this fairground near the beach with all the usual fairground bollocks. You know, coconut tosses, bumper cars, cotton candy, that sort of stuff. It ain't there no more, but it used to be quite popular, back in the day."

He pulled the umbrella from his cocktail and licked it clean before using the end as a toothpick.

"The place was owned by Harry McGuire, right nasty piece of work he was. Big Gypsy bastard, you know the type, mouth full of gold and fingers full of rings. Rings very much like these, come to think of it." He held up his hands to reveal his brash collection of signets, sovereigns, and affiliations, one of which had Harry McGuire's initials etched into it.

Danny pretended to admire the mobile museum of ill-gotten gains.

"I rented a plot off old Harry McG, a prime piece of real estate between the bumper cars and the carousel. It was right near the entrance, so everybody had to walk past it on their way in and out of the fairground, which was good news for me but bad news for the parents, 'cos there ain't a kid in the world who can walk past a bouncy castle without wanting to kick their shoes off. The plot cost me an arm and a leg, but I was raking in so much dough in that first month that I could have paid for the next year up front if only I'd been smart about it. Being a daft sod in my twenties with more money than sense, however, I did what any daft sod in their twenties would do and I spent the fucking lot of it. June of seventy-four was a hell of a month, Dan, let me tell you. Never had one like it, before or since."

Reg swirled his straw around and smiled into his drink as if he'd just stirred up some long-forgotten memories.

"Course, I wasn't to know that the shittiest summer since the dawn of time was just around the corner, otherwise I might have put a little bit aside for a rainy day, or, as it turned out, a whole fucking season. The Dreaded Dresden Drizzle, that's what the papers called it. Fuck knows why Dresden. I think they just wanted to blame Germany. Hating the krauts was still pretty hip back in those days. You couldn't even call a German shepherd a German shepherd without looking like some kind of Nazi sympathizer, you had to call it an Alsatian, as if sympathizing with the French was any better. Anyway, whatever. June was a scorcher, but then July comes around and it's wetter than a beaver's sweater. Still, not to worry, I thought, it'll clear up eventually. But four weeks pass and suddenly it's August and it's still fucking pouring down, and then before you know it we're into September and guess what?"

"Still raining?" said Danny.

"It's still raining, and it don't stop until we're well into autumn, by which time I'm up to my bollocks in debt. See, most of the other attractions were weatherproofed—some had roofs, others had tarps—but the Boogie Bounce had an open top, so instead of a bouncy castle I ended up with what looked like a giant hook-a-duck pond, but one with real ducks floating around in it. Days went by without a single customer. Well, apart from this one lad, Ricky he was called. Odd little fucker he was, used to bite people like a rabid monkey. One time I caught him trying to bite a hole in the Boogie Bounce, I had to keep smacking him around the head with his own shoe until he let go."

He fished the cherry from the glass and threw it into his mouth. Danny winced when he heard a crunch, but Reg simply chewed through the pit the way a crocodile chewed through a bone. Mo had once told him that cherry stones contained a compound that the human body turned into cyanide, but Danny chose not to share this information with Reg.

"Sorry, where was I?" he said.

"Rabid monkey child?" said Danny.

"Before that. Oh yeah. So, every week Harry would come for his rent, and every week I'd give him what I had and tell him I'd make up the rest the following week, but I never could, and the fact that Harry kept adding interest didn't make life any easier. So the debt kept mounting up, and all this time I'm losing money 'cos I'm not getting any customers, and all this time *he's* losing money 'cos I can't pay the rent. It wasn't my fault and it wasn't Harry's either. It wasn't even the fucking Germans' fault. It was just shit luck, plain and simple. But still, it wasn't exactly what you'd call a sustainable arrangement, so I wasn't surprised when Harry decided to kick me off the fairground. I wasn't even surprised when he told me he was keeping the Boogie Bounce as collateral. It was worth at least three times whatever I owed him, but I didn't care by that stage, I just wanted out, so I said he could have the cursed thing. What *did* surprise me, however, and it really fucking shouldn't have, was that Harry decided the Boogie Bounce wasn't worth even half what I paid for it, which meant I still owed him a shit-ton of money that I couldn't afford to pay, and he knew I couldn't afford to pay it, which is why he did it in the first place, because if there's one thing that Harry enjoyed more than making money, it was breaking bones. You ever seen *Misery*, Dan?"

Danny nodded.

"Well, it was a bit like *Misery*, but instead of Kathy Bates it was Harry, and instead of being tied to a bed I was held down on a pool table by four of Harry's sons while seven of 'em watched. I dunno where the other three were. Harry took a lump hammer that they used to knock in the marquee pegs and he didn't stop swinging until my legs were like two bags of broken biscuits. Then, and only then, did he agree to call it square."

Reg drew on his straw until it gurgled against the bottom of the glass.

"That horrible old bastard taught me a valuable lesson that day. A painful lesson, mind you, but a valuable one. I learned that sometimes, whether you like it or not, you've just got to pay the price, even if you ain't done anything wrong. Sometimes things happen and they're completely out of your hands, like when it rains all fucking summer when it ain't supposed to, and it ain't right, and it ain't fair, but you've still got to pay the price. You understand what I'm saying, Dan?"

"Yes, Reg," said Danny, wiping his clammy hands on his trousers.

"Good lad," said Reg. "I knew you would. Dent?"

Mr. Dent passed Reg his crutches and helped him off his stool.

"Enjoy your drink," Reg said on his way out. "Don't forget to tip Charlie when you pay for 'em."

CHAPTER 14

An elderly lady with a tartan shopping trolley and frilly bedroom slippers on watched as Danny struggled to find the enthusiasm to dance for such a pitiful crowd. By her feet lay a tatty black schnauzer, who hadn't blinked for so long that Danny was concerned it might have died, perhaps in shock at his performance. When the song was over, he took a bow and waited for some kind of reaction, preferably the kind where she gave him money, but the lady didn't flinch and neither did her dog. Nor did they react when he accidentally on purpose shook his money box at them by accidentally on purpose giving it a kick. He stood there awkwardly, like somebody who had farted in a crowded lift that wasn't stopping for another ten floors, and hoped the awkwardness would spur her into action, but the lady seemed completely at home with awkwardness, which only made him feel more awkward, and so he kept on dancing. This had been going on for forty-five minutes, and it was almost an hour before the lady finally reached into her handbag, fished out her purse, extracted two chocolate limes and a dirty deutsche mark, and dropped them into Danny's lunchbox. He thanked her and watched as she shuffled across the park with her trolley trundling behind her, seemingly unaware that her misty-eyed dog was still curled up where she'd left it. Only when Danny poked it with his toe did the animal lazily rise from the dead and stagger off after its owner.

He unwrapped a chocolate lime and popped it into his mouth. Slipping the wrapper into his pocket, he felt something else in there and removed his hand to find one of the napkins that Krystal had given him. He stuffed it back into his pocket and swilled his lunchbox around to reveal another paltry day's takings.

Not for the first time that week, Danny caught himself seriously questioning his life choices. It wasn't even the first time that *day*. He already found it difficult enough to look at himself in the mirror, but he was starting to find it even harder when all he saw was a hopeless panda staring back at him. The sole purpose of buying the costume was to make some quick and easy money, but he was no closer to paying off Reg than he had been three weeks ago. Nor was he any closer to knowing how to dance. Whatever brain glitch had led him to believe this was a good idea had long since resolved itself, and Danny could now see his situation for precisely what it was.

"Ridiculous," he muttered as he stared at the handful of coins in his lunchbox.

He sighed and checked the time. It was late afternoon and the park was emptying of people, including the performers who were busy packing up after another comparatively lucrative day. The one-man band was heading for the exit with his cymbals still attached to his knees, confirming Danny's suspicions that the man was actually deaf and not just a terrible musician; the human statue was literally running out of the park, as if trying to compensate for all of those hours spent standing still; and El Magnifico was grinning to himself as he leafed through an offensively large bundle of money. Danny prayed for a sudden gust of wind to blow the notes from his hands and into the blades of the industrial lawnmower that was currently trundling around the park.

Tim approached with his guitar on his back and Milton draped around his neck like a furry boa.

"How's the life of a dancing panda?"

"See for yourself," said Danny, nodding at his lunchbox.

"Oh," said Tim when he saw the contents. "Wait, is that a chocolate lime in there?"

"Yep. If only sweets, bottle caps, buttons, and rocks were legal tender, I'd have my rent paid off in no time."

"Trade you for a Starburst."

"What color?" said Danny.

Tim fished around in his pocket. "Red," he said, inspecting the sweet.

"Deal."

"I love chocolate limes," said Tim as he unwrapped the sweet and popped it into his mouth. "They remind me of my gran. She was like a chocolate lime dispenser."

"Yeah, well, they remind *me* of what a stupid idea this whole thing was."

"Hey, I was performing for weeks before somebody gave me my first chocolate lime. Stay positive, you're making progress."

"Not fast enough," said Danny, closing his lunchbox with a sigh and stuffing it into his bag. "How long have you been doing this anyway?"

"Four years. I used to come here to practice whenever my housemates at uni got sick of the noise. I wasn't planning on making a living out of it or anything, but, well, I wasn't planning on being adopted by a cat either."

"You two seem like a pretty good team," said Danny, nodding at Milton.

"Team?" said Tim. He laughed. "This isn't a team, this is a prolonged hostage crisis. One day I found him asleep in my guitar case and he's followed me around ever since."

"What were you studying at uni? Music?"

"Finance."

Danny laughed. "Seriously?" he said.

"What?"

"Nothing. I just . . . you don't exactly look like the finance type."

"I'll take that as a compliment," said Tim. "I wanted to be an investment banker, believe it or not. Correction: my *mum* wanted me to be an investment banker and, well, she was paying for my education so I didn't really have much say in the matter."

"I bet she wasn't too happy when you dropped out, then," said Danny.

"You think I'm a dropout just because I play a guitar and have a cat on my shoulder?"

"No, I think you're a dropout because you're literally wearing a badge that says DROPOUT," said Danny. He pointed to a cluster of colorful pin buttons that decorated Tim's breast pocket.

"Oh. Yeah. I just bought that because it seemed edgier than a badge that said MASTER OF FINANCE on it."

"So you graduated?"

"Top of my class."

"Then no offense or anything, but what the hell are you doing here? You could be raking it in, mate, you don't need to be doing this."

"Most of us don't *need* to be doing this. We do it because we enjoy it. This isn't the Foreign Legion, you know. People don't just become street performers because they're in some kind of trouble. Well, apart from you. No offense or anything." Tim smiled.

"*Touché,*" said Danny.

"I tried it for a couple of years. The banking thing. I hated it. The money was good but I was totally miserable, and so was everybody I worked with. Whoever said that money can buy you happiness clearly had no idea what they were on about."

"Nobody said that," said Danny.

"What?"

"Nobody ever said money can buy happiness. They said money *can't* buy happiness."

"Really?"

"Pretty sure, yeah."

"Then my mum's full of shit," said Tim.

"Well, money seems to make him happy," said Danny, nodding at El Magnifico, who was busy counting his bills for the umpteenth time.

"It's basically the only thing he cares about. Well, that and his gown. He's very attached to the gown. Some kid accidentally stepped on it once and he threatened to fucking destroy them. Those were his actual words. The mother was quite rightly mortified."

"It's not even a gown, it's a woman's bathrobe. He stole it from his ex-girlfriend."

"Are you sure?"

"That's what I heard."

"No, I mean are you sure he has an ex-girlfriend? Like, a real one?"

"Oh, she's real all right," said Danny. "Real mean. Incredible dancer, though, I've never seen anything like it. I asked her to teach me but, well, she called me a twat and gave me the finger. And then robbed me."

"Maybe she'd be more willing to help if you came bearing gifts."

"Like what? I can't even afford the bus ride home."

"How about a nice silk bathrobe?" said Tim, his eyes fixed on El Magnifico, who had finally stopped counting his money and was now carefully folding up his gown. They both watched him slide the garment into his bag.

"We can't," said Danny.

"Why?"

"Because. It's stealing."

"You can't steal something that's already stolen. It's like, you know, double jeopardy or whatever. And anyway, think of it as payback for stealing your clothes."

"We don't know for a fact that was him."

"The human statue saw him do it. She told me."

"And she didn't try to stop him?"

"She didn't want to break character."

"Brilliant," said Danny. He chewed his lip and stared at El Magnifico. "Okay, how are we going to do this?"

Tim grinned.

"Follow me," he said.

El Magnifico was busy emptying his sleeves of flowers, playing cards, and colorful strings of handkerchiefs. He didn't see Tim approaching.

"Milton's written a song for you," said the musician. "Want to hear it?"

El Magnifico ignored him.

"Great," said Tim, strumming his guitar and twiddling the tuning pegs until he found the sound he wanted. He cleared his throat and began to sing in the style of a medieval ballad.

"There once lived a rubbish wizard, his gaze was icy like a blizzard, his face was ugly, like a lizard, his name was El Magnifico."

"Go away, hippie," said El Magnifico without turning around.

"He liked to dress in women's clothes, he wore a bathrobe to his shows, beneath it he wore pantyhose, and not just at the weekend."

"I'm warning you," said El Magnifico, unaware that Danny had crept up behind him as he turned to face Tim. "Leave now, while you're still not on fire."

"He thought he could set things alight, with nothing but his mental might, his face would go all red and bright, but nothing ever happened."

"Right!" said El Magnifico. "You've asked for it!" He pointed at Milton. "Say good-bye to your little friend!" He placed his fingers on his temples and grimaced like he'd stubbed his toe. He didn't see Danny gently tugging the robe from his bag.

"Then one day his head went bang! 'Thank God for that!' the people sang, flowers bloomed and church bells rang, and world peace shortly followed."

El Magnifico started to tremble like the back row of an adult movie theater. Behind him, Danny gave Tim the thumbs-up and tiptoed away with the robe under his arm. Tim nodded and began to walk off in the opposite direction.

"That's the ending of this tale, now it's time for us to bail, 'cos H&M are having a sale, and Milton wants a turtleneck sweater."

The magician gasped, his body slumping as his hands fell by his sides.

"Next time, hippie!" he shouted after Tim. "Next time!"

CHAPTER 15

Litter swirled around Danny's feet as he walked through a part of town that was always dark, regardless of the weather or time of day. A drunk man staggered into him and blamed him for it as he zigzagged down the middle of the road like a sailor crossing a deck in a storm. Even the pigeons were hostile, standing their ground like feathered thugs and forcing Danny to walk around them.

He stopped outside a pair of black double doors. Above them hung the transparent skeleton of an unlit neon sign that read FANNY'S. Not wanting to be seen entering a strip club at barely ten o'clock in the morning, Danny lingered outside for a moment until the street was clear of people. Then, trying one of the handles and finding the door unlocked, he cautiously entered the club.

He found himself in a long dark corridor that smelled like baby wipes and something else that he couldn't identify but guessed to be the odor that emanated from broken dreams. Passing an unmanned coat booth and then the bathrooms—the men's bore a sign reading DICKS on the door while the women's read LADIES—he entered a large room containing several empty podiums and more mirrors than a Borges anthology. Each of the platforms had a pole in the middle that reached up to the treacle-colored ceiling tiles that seemed to have been installed at least a decade before the smoking ban came into effect.

At the far end of the room was a bar, and at the bar was a woman dressed like a judge from a children's beauty pageant.

"Hi," said Danny.

"Fuck off," said the woman without looking up from her calculator. "We're closed."

"I'm looking for Krystal."

"So is everybody. Come back tonight like everybody else. Until then, fuck off."

"Look, if I could just see her for a minute, I've got something I think she might like."

"That's what they all say."

"Really?" said Danny.

The woman sighed and looked at him then. She was caked in makeup, but no amount of foundation could dilute the lifetime of heartache, disappointment, late-night phone calls, and early morning pick-me-ups that lived in the creases of her face.

"Yes, fuckstick, really, and you know what? Somebody out there in this miserable world of ours probably *is* interested in what you have to offer. You might have to pay them, or get them really drunk, or dig them up, or order them online, but still, they're out there, somewhere, waiting for you, so turn around, walk out the door, and go find that special someone, because I guarantee you that Krystal has absolutely no interest whatsoever in seeing your shriveled excuse for a penis. Got that?"

"My shriveled . . . wait, what?" said Danny. "No, that's not— Let me show you—"

"Vesuvius!" yelled the woman.

Before he could ask what exactly a Vesuvius was, a man burst through a door behind the bar with arms full of muscles and muscles full of tattoos.

"This weirdo was about to get his bits out," she said.

"I was not about to get my bits out!" said Danny, unable to believe he was having this conversation.

"Do us a favor and show him the door, would you, Suvi?"

"Can I frisk him first?" said Vesuvius. "He looks like he'd enjoy it."

Danny looked at the man's knuckles. OPEN was written across his right hand, WIDE across his left. Danny had no idea what it meant, but every explanation he could think of sounded painful.

"It's okay, Fanny," said Krystal as she suddenly appeared behind Danny. "He's harmless. He's a dickhead, but he's harmless."

"That's right," said Danny, nodding. "I am. Harmless, I mean."

"Then I guess it's your lucky day," said Fanny. "Come on, Suvi, let's leave these two lovebirds alone."

Vesuvius, clearly disappointed with the nonviolent resolution, skulked after Fanny as she disappeared through the door behind the bar.

"Thanks," said Danny. "I—"

"You got that money you owe me?"

"What? No, I—"

"Then fuck off," said Krystal as she turned to leave.

"Wait!" said Danny, digging around in the bag he was carrying. "I brought you this." He pulled out El Magnifico's robe and held it up.

Krystal stared at the garment. She tried to frown, but her mouth twitched with the urge to smile. "Where'd you get that?" she said.

"Magic."

Krystal started to laugh. Danny smiled.

"What's so funny?" he said.

"It's not mine."

"What?"

"The robe. It's not mine. I was just saying that to piss him off," she said.

Danny shrugged. "Well, I guess this'll piss him off even more, then."

She took the robe and turned it over in her hands. "Why'd you do it?"

"Because I'm a good person," he said. Krystal stared at him. "And because I need your help."

"I fucking knew it!"

"Just teach me the basics," said Danny. "Please. Just the basics. That's all I'm asking."

"No," said Krystal.

"Please."

"No."

"I'll give you a hundred pounds," he said, trying and failing to provoke a smile.

"I'll give *you* a hundred pounds to jump in front of a bus."

Danny thought about this for a moment.

"Does the bus have to be moving?" he said.

"Yes."

"How fast?"

"Fast enough to kill you," said Krystal, "but slow enough so you'll feel it."

"Right," said Danny. "That's not a good deal."

"Yeah, and neither is me wasting my time teaching you how to dance in exchange for sweet fuck-all."

"Look, think of it like this. The better I can dance, the more money I can make, and the more money I can make, the more I'll be taking away from El Magnifico. Don't think of it like you're helping me. Think of it like you're screwing your ex. Figuratively speaking."

This time Krystal said nothing. She chewed her lip and shook her head as if disagreeing with some invisible counsel.

"Please. Just do this and you'll never have to see me again."

"You promise?"

"Cross my heart and hope to die."

"That makes two of us," said Krystal. "Come on, then." She gestured for him to follow her. "Let's get this over with."

"Wait, now?" he said, but Krystal had already gone.

Danny followed her through the door behind the bar, down a short corridor, and into a large room with wooden floors and a grubby mirror lining the full length of the back wall. It looked like a ballet studio, albeit one with disco balls and pole-dancing equipment, and it smelled like cigarettes and Red Bull.

"You've got two hours," said Krystal. "If you can't learn the basics in two hours, then you may as well go back to stacking beans or manning glory holes or whatever it was you were doing before."

"I was a builder, actually. I lost my job about a month ago."

"Didn't ask, don't care," she said, crouching in front of a large stereo system in the corner of the room. "Let's get started, I haven't got all day."

The mirror began to vibrate as ABBA's "Gimme! Gimme! Gimme!" started pounding from the speakers.

Krystal hung her coat up and stood in the middle of the room. Danny lingered nervously by the door.

"Come on, then, numb nuts," she shouted, pointing to the spot beside her.

Danny took a deep breath and joined her in front of the mirror.

"Okay," said Krystal. "Stand like this. Feet apart, head down, wait for the beat to kick in. Three, two, one, nowwwww start with the shoulder. Nice and easy, keep it loose."

She gently rocked her shoulder as she moved in time to the music. Danny jerked his arm back and forth like a faulty factory robot.

"Now the other shoulder," she said. "Like this. Left, right, left, right. Just follow the rhythm and go for it."

Danny followed the rhythm and went for it, but the rhythm saw him coming and ran off before he got there.

"Then slowly bring the hips in, and then the arms, like this. Small movements, nothing fancy. Click your fingers if it helps. Like this. Click. Move. Click. Move."

He started to snap his fingers, but the gesture only confused him further. He looked like he'd danced into the Twilight Zone and didn't know how to dance his way out again.

"Now the feet. Blokes never use their feet, they're too scared to spill their pint, but you can't dance without your feet. Again, keep it simple, like this. Step one, step two, step one, step two."

Danny wiped his brow with his forearm. He felt like he'd just started a marathon and only now realized how far away the finish line was.

"Come on," said Krystal, "keep it up. You're doing well. Now bring it all together. Head, arms, shoulders, hips, legs. Feel it. Come on. Dance. Dance like you want a man after midnight."

"Head, shoulders, arms, head, shoulders," wheezed Danny as he moved the opposite part of his body to every part he muttered aloud. His face glistened like the disco ball above him, whether with sweat or with tears, he wasn't quite sure.

"Last stretch. Don't lose it now. Dance. Twenty seconds. Give it all you've got. Ten seconds. Come on. Five seconds. Four. Three. Two. One. And. Rest."

Danny slouched forward, steadying himself with his hands on his knees to keep from keeling over. Sweat dripped from the end of his nose and his breathing was loud and labored. He felt like he was about to throw up, or die, or throw up and then die.

Krystal smiled at him in the same way a murderous spouse smiles at her partner just before skydiving together.

"Ready for round two?" she said.

Danny was no stranger to suffering. He was, if anything, a very close acquaintance of it. But over the course of the next two hours, all of the problems that had plagued his life up until that moment miraculously disappeared, not because he was lost in the music or the art of the dance but because keeping up with Krystal was so traumatic that all other traumas took a backseat while he focused on simply trying to survive.

It wasn't easy for him to pinpoint precisely which part of the process he struggled with the most because he struggled with everything equally. Fitness proved to be a major obstacle. Danny had always considered himself to be a fairly healthy guy. He couldn't run a marathon, or even any kind of extended distance unless being chased by something ravenous, but he could sprint for a bus without risking an aneurysm and he could take the stairs if the elevator was broken without first letting the police know where to find his body. He didn't eat organic kale with tofu for breakfast every day (nor any day, for that matter), but he didn't smoke, he rarely drank, and while years in the construction trade had turned many of his workmates into redder, heavier versions of their former selves due to the self-deceptive belief that routinely eating pastries was fine as long as you kept yourself moving, the daily grind of the building site had transformed the scrawny kid that Danny had started out as into the strong, lean man he now was.

But early into the session it became painfully clear that it wasn't strength he was lacking, nor was it strength that he needed (except for strength of mind, perhaps, his having absconded almost as soon as the session had started); it was stamina. He could barely make it through a single song without pausing in the middle to catch his breath, check his pulse, and google how many beats per minute it took for a human heart to explode, and even without the panda suit he perspired so profusely that at one point Krystal had to call the cleaning lady to squeegee the floor lest it turn into a safety hazard. Equally problematic was Danny's coordination skills, or lack thereof. He shook when he should have been shimmying, he shimmied when he should have been spinning, he spun when he should have been strutting, and instead of strutting he did something that even Krystal didn't have a word for. Not that she made it particularly easy for him. Following her lead was like following a fugitive who knew the roads when he didn't. She gunned the straights, sped into corners, and only slowed down when Danny took a wrong turn or ended up in a ditch. Even when she took her foot off the pedal he struggled to keep up with her, and so it went for two exhausting hours until Krystal finally stopped the music and threw him an unwashed beer towel to wipe his face with, which he did, gladly. She looked impossibly composed, like someone who had just awoken from a long and invigorating sleep, and the only time she broke a sweat was when she had to assist Danny out of the studio and back down the corridor to the bar.

"Two waters, please, Suvi," said Krystal, "and a kiss of life for this one." She nodded at Danny, who was trying to hoist himself onto the barstool beside her.

"I won't charge for the kiss," said Vesuvius, winking at Danny as he placed two bottles of water on the counter.

"Don't worry," said Krystal as Vesuvius went back to cleaning glasses. "He only likes the married ones."

"I *am* married," said Danny in between deep gulps of water. "Sort of."

"Sort of?"

"It's a long story," he said, looking at his empty ring finger. He'd never worried about his wedding ring when Liz was alive, but after her death he suddenly became terrified that he was going to lose it, so he'd wrapped the ring in cotton, put it in a matchbox, and hidden it in the drawer of his bedside table, which was where it had lived ever since.

"Every man who comes in here has a long story," said Krystal. "Don't tell me, let me guess. Your wife ran off with another bloke? We get a lot of those."

Danny shook his head and took another sip of water. Krystal thought for a moment.

"She ran off with another woman?"

"Nope."

"Another panda?"

"Funny."

"Was it a dwarf? Because we had this one guy whose wife—"

"She's dead," said Danny.

Krystal watched him for a moment, her lips twitching with a wavering smile. "You're joking, right?" she said.

"I wish I was," said Danny, screwing the cap back onto the bottle.

"What happened?"

"She died in a car crash, just over a year ago."

"Shit," said Krystal. "Sorry." She spun her bottle top between her fingers and watched it dance across the bar. "My mum always said I had a big mouth."

"Sounds like I'd get on well with your mum," said Danny. He smiled.

"You'd be the only person who does," said Krystal. She took a sip of water, and the two sat in silence for a minute.

"She was a dancer, actually," said Danny. Krystal frowned. "My wife. Liz."

"She clearly never taught you much."

"No shit," said Danny. "Thanks for today, by the way. I know you didn't want to do it, but I'll probably die in my sleep tonight if it's any consolation."

"I thought you were going to croak back there, to be honest."

"I was close. Everything started to flicker at one point and I remember thinking, *So, this is it. This is how it ends.*"

"I keep telling Fanny to get that strip light fixed."

They both smiled.

"Seriously, though, thanks," he said.

"Don't mention it. I never thought watching a grown man on the verge of throwing up for two straight hours could be so entertaining."

"I really don't know how you do it," said Danny, wincing as he gently massaged his knee.

"It's easy. You get up early, you go to bed late, you always make sure you have ice in the freezer, and you go through so much Voltarol that you end up with a borderline diclofenac addiction. You dance six or seven days per week for four or five hours per day, month in, month out, for about five years, and voilà, you too will be qualified to spin around a pole in a dodgy nightclub while crusty old wankers stuff clammy ten-pound notes in your underwear."

"The way I dance, I'd be lucky if people stuffed ten-pence pieces in my underwear."

"You weren't that bad, to be honest. I mean, don't get me wrong, you were terrible. Like, *really* fucking terrible. Like, so bad that I actually started to feel sorry for you. But still, you weren't as bad as I expected. Nothing a little practice won't fix anyway. Okay, a lot of practice. A metric shit-ton of practice. But you'll get there."

"Not until I find some rhythm, I won't. I've got more chance of catching a bullet in my teeth than I do of catching a beat."

"Now *that's* something I could help you with," said Krystal. She sounded sincere.

"I bet," said Danny.

"Get yourself a metronome."

"A what?"

"It's this thing with a hand that ticks," said Krystal. Danny followed her finger as she wagged it for emphasis. "You use it to keep time."

"Isn't that called a clock?"

"No, you muppet, a clock is— I'm not even going to explain what a fucking clock is. Just google it. Metronome. Get yourself one. And use it. Not just when you're trying to dance, but whenever you're doing anything at all. Chopping onions. Washing the dishes. Brushing your teeth. Cleaning the windows. Do it all to the metronome, and you'll learn to keep a beat without even knowing it."

"Nothing's coming up when I search for it," he said. He showed his phone to Krystal.

"Not 'metro gnome,' you fucking— Jesus, give it here." She snatched the phone from his hand and tapped away at the screen, her nails clacking against the glass. "That," she said, handing the phone back to Danny. "Buy one. Or even better, download one for free. Just search for metronome apps."

"Thanks," said Danny. "Any other tips?"

"Yeah, rewatch all the dancing films you've ever watched, and then make a list of all the ones you haven't and watch them too."

"That's going to be a very long list."

"You've never seen a dancing film?"

"It depends on what you call a dancing film."

"Like a film, but about dancing."

"Oh. No, then."

"*Flashdance*? *Footloose*? *Billy Elliot*? *Strictly Ballroom*? Please tell me you've at least seen *Dirty Dancing*. Tell me that and I can just about forgive you," she said.

Danny shook his head. "Liz was always trying to get me to watch it with her," he said, suddenly unable to think of a single decent reason why he never had. "It was one of her favorites."

"She sounds like my kind of girl," said Krystal.

"And mine," he said. He turned his water bottle upside down and watched the droplets zigzag down the plastic.

"Well, watch it. And then rewatch it. Like a hundred times. Everything you need to know is in that film. Not just about dancing but about life." Krystal slid from her stool. "Anyway, I better go warm up."

"Warm up?" said Danny. "How are you not warm? You've just been dancing for two hours."

"Different kind of warm-up," said Krystal. She nodded towards one of the podiums where another dancer was circling a pole with a cigarette pinched between her lips.

"Got it," said Danny, taking the hint. "I'll leave you to it."

"Good luck with the whole panda thing," she said over her shoulder.

"Thanks," he said, fully aware that he was going to need it.

CHAPTER 16

Will watched Mr. Coleman shuffle the length of the whiteboard, his marker squeaking against the enamel as he scribbled something in capital letters.

"International 'What Do Your Parents Do for a Living' Day," said Mr. Coleman, reading the words he'd just written down. "Anybody ever heard of this?"

Some of the kids shook their heads. Others stared blankly.

"Well, me neither, but apparently it's a thing, and apparently we have to talk about it today. By the way, for those of you who were wondering, yesterday was International Duck Day—I know, I can't believe we missed that one either—and tomorrow is International 'Teachers Are Amazing and Rarely Appreciated and Are Overworked and Underpaid' Day, so make sure you spread the word."

The only sound in the room came from somebody's pencil rolling off the desk and landing on the floor.

"So the point of today is to celebrate, well, capitalism, I imagine, although according to this handy fact sheet that the Education Gods were kind enough to furnish me with, it is, and I quote, 'a day to celebrate the many and diverse ways in which our parents help to keep the world turning.' Your parents don't keep the world turning, just so you know. *Physics* keeps the world turning, but you get the idea. For those of you whose parents are not currently employed, fear not, the world will not suddenly grind to a halt. It will continue to spin, at least until the day when the sun implodes and vaporizes this sorry little planet we call home. For those of you whose parents *are* currently employed, however, how many of you know what they do for a living?"

A number of kids raised their hands, including Will, who still had no idea about his dad's recent change in circumstances.

"Great," said Mr. Coleman. "How's about some of you come up here and tell us all a little bit about their work. Mum or dad, it's up to you. We don't need a monologue, just a few words."

Mr. Coleman saw Will's hand slinking beneath the sea of arms.

"Actually," he said, "you know what? Let's play a game, shall we? Instead of *telling* us what your parents do, how about you *show* us."

"What, like a video?" said Jindal.

"Why would I have a video of my mum at work?" said Atkins.

"I've got one of my dad battering a shoplifter when he worked security at Zara," said Kabiga as he scrolled through his phone.

"I'm not talking about videos," said Mr. Coleman. "I'm talking about using your imagination!"

A low murmur of discontent passed through the class at the mention of the word *imagination*.

"Like charades?" said Mo.

"Precisely, Mo," said Mr. Coleman. "Like charades."

"Why can't we just say it?" said a kid at the back of the class.

"Because that's boring! That's what all the other classes are doing. This will be more fun." Mr. Coleman glanced at Will, who gave him the faintest smile of acknowledgment.

"You got a weird idea of fun, Mr. C."

"Look, I'll go first, okay?" said Mr. Coleman. "I'll show you what my dad used to do for a living."

He took off his jacket and draped it over the back of his chair.

"Your dad was a stripper?" shouted somebody. Laughter filled the classroom.

"Very funny," said Mr. Coleman. "I haven't started yet. Okay, here we go."

He sat down on the chair and picked up a pair of invisible drumsticks. Then, clacking them together above his head like a Monsters of Rock headliner, he launched into an unexpectedly enthusiastic albeit entirely silent drum solo.

"Guitarist!" shouted Cartwright.

"Drummer!" shouted everybody else.

"Exactly!" said Mr. Coleman, wiping the sweat from his brow.

"Same thing," muttered Cartwright.

"A percussionist for the London Symphony Orchestra, to be precise. You should have seen him, he was quite the player."

"I bet he was," said Mo. "Girls always fancy the drummer."

"How do you know?" said Kabiga.

"Because he *is* a girl," said Claire Wilkins, who sat behind Mo. Everybody sniggered.

"That makes one of us," said Mo. The room erupted with chants of "Burn!" and "Savage!"

"I didn't mean *that* kind of player," said Mr. Coleman, belatedly realizing his slipup. "I meant player of the *drums*. Although he was quite popular with the ladies, actually. Not quite so popular with my mum though. Anyway, who wants to go next? Mo? Come on, show everybody how it's done."

"Easy," said Mo. He took a seat in Mr. Coleman's chair, held an imaginary steering wheel in front of him, and proceeded to honk his horn, yell in Punjabi, and flip off the various other drivers committing multiple infractions on whatever imaginary road he was driving on.

"Taxi driver!" yelled the class.

"Well done, Mo," said Mr. Coleman as Mo returned to his seat. "Please remind me never to get into your dad's taxi."

"My dad's an estate agent," said Mo. "My mum's the taxi driver."

"Then please remind me to say wonderful things about you at parents' evening. Will, your turn. Get up here and show us what you've got."

Will shuffled to the front of the class. He stood around looking sheepish for a minute before grabbing an imaginary shovel and halfheartedly driving it into the ground.

"Coal miner!" shouted somebody.

"Gold digger!" shouted another kid.

"Drummer!" shouted Cartwright.

Everybody laughed, including Will. He put down his invisible shovel and started laying bricks instead, this time with more enthusiasm, but he looked more like he was scaling a wall than attempting to build one.

"Rock climber!"

"Spider-Man!"

Will, laughing even more now, turned to Mr. Coleman for help. The teacher smiled and shrugged.

"Don't look at me, Spider-Man!"

Will picked up an imaginary hammer and pretended to bash it into a nail.

"Handyman!" shouted Jindal.

Will pointed at Jindal and prompted him to elaborate.

"Carpenter!" shouted Jindal.

"Builder!" shouted Cartwright.

Will pointed at Cartwright and gave him the thumbs-up.

"Well done, Will. And well done, Cartwright!" Cartwright beamed like he'd just got a C-minus in a maths test. "Take a seat, Will, I think you need a rest after that."

Some of the kids patted Will on the back or playfully thumped him in the arm as he made his way to his desk.

"Right," said Mr. Coleman, exchanging a nod with Will. "Who's next?"

The game went through a few more rounds (there would have been more, but one kid spent at least ten awkward minutes trying to find a way of showing the class that his mother was a gastrointestinal endoscopist) before the bell rang and everybody commenced that most paradoxical of migrations, the one where students rush from one class and drag their heels to the next.

"Will?" said Mr. Coleman as Will and Mo tried to fit through the door at the same time. "Can I have a quick word?"

The two boys shared a look of dread before Will returned to the classroom to face whatever punishment was coming for whatever it was he'd done wrong.

"It's okay, Mo," said Mr. Coleman when he saw Mo lingering in the doorway. "I won't be needing your mediation services today, thank you."

Mo looked at Will, shrugged, and closed the door behind him. Will took a seat.

"Nice bit of acting back there," said Mr. Coleman. "You could be a movie star. Well, a silent movie star anyway."

Will smiled. He stopped fidgeting and waited for Mr. Coleman to find the right words for what he wanted to say next.

"I know this is none of my business, Will, and I'm sure you're probably sick of people giving you advice or telling you what to do, but, well, I just want you to know that I get it. The silence, I mean. I can't pretend to understand what you're feeling, or what you've gone through this last year, but I do know a little bit about what it's like to lose somebody close to you."

Will looked down at his hands and gently picked at the corner of his thumbnail.

"My grandfather died when I was about your age. He was more like a parent than a grandparent though. My dad was always busy with orchestra rehearsals and my mum was a nurse who often worked nights, so my grandfather basically raised me for the first ten years of my life. His hair was gray and his eyes were gray and everything in his wardrobe was gray, but when he walked into a room it was like the sunshine had walked in right behind him. He was old even then, but I still thought he'd be around forever, because, well, you do, don't you? So when he died it came as a massive shock, like a train had come out of nowhere and slammed right into me. I didn't speak about him for a long time afterwards. Not to my mum, not to my dad, not to my friends. I just didn't know what to say. Talking about him in the past tense seemed so strange that I couldn't bring myself to talk about him at all, if that makes any sense."

Will nodded, his eyes still fixed on his hands.

"One day my dad gave me this old stuffed rabbit toy. I'd never seen it before, but he said it belonged to my grandfather. He'd found it while clearing out his belongings and he thought I might like to have it. The rabbit was called Colin, and it was the saddest thing you've ever seen. Three limbs, one ear, big clumps of fur missing. He looked like he'd been run over by a lawnmower. In fact, I think he actually *had* been run over by a lawnmower. But Colin made the best listener, even with one ear. I could talk to him about my grandfather in a way that I couldn't talk to people."

The corner of Will's mouth twitched slightly.

"I know, I know. Go ahead and laugh. I'm only telling you this because I know you won't talk. Otherwise I'd probably be out of a job tomorrow."

Will reassured Mr. Coleman that his secret was safe with a finger to the lips and a subtle nod.

"Thanks, I appreciate it," said Mr. Coleman. His chair groaned slightly as he leaned back and crossed his arms. "The truth is that I felt comfortable talking to that rabbit because, unlike everybody else around me—my family, my friends, my teachers—Colin wasn't trying to fix me. He didn't pretend to know how I felt. He didn't expect me to 'be like I was before'"—Mr. Coleman made quotation marks with his fingers—"as if nothing had changed and life should somehow continue as normal despite this gaping hole that had suddenly appeared right in the middle of it. He didn't expect anything because, well, he was just a stuffed toy. All he could do was listen. So that's what he did. He listened. And it helped. Before then I didn't think I'd ever be able to talk about my grandfather again. But I've since come to realize that difficult things don't necessarily have to be difficult to talk about. The difficult part is finding the right person—or rabbit—to talk to. Not that I'm saying you should start talking to animals, of course, although if you do, then please, don't tell anybody that I was the one who suggested it. I'm sure Colin would be happy to give you a free consultation, though, and you're more than welcome to talk to me anytime, even though you probably think I'm crazy by now. We can play charades if you'd prefer."

Will smiled as he pictured Mo yelling at the class from his imaginary taxi.

"I guess what I'm trying to say, Will, is that when terrible things happen that we can't understand, sometimes it takes something equally unexpected to help us make sense of it. Do you see what I'm saying? Or am I rambling like an idiot?"

Will rocked his head from side to side to indicate a bit of both.

"Okay, that's good enough for me," said Mr. Coleman. He tore a piece of paper from a spiral notepad, scribbled *Sorry Will and Mo were late, I was rambling like an idiot*, and signed it before handing it to Will. "Give this to your next teacher so you don't get into trouble."

Will looked at the note and frowned when he saw Mo's name. Mr. Coleman nodded towards the door behind Will, who turned just in time to see Mo's face quickly disappear from the window.

"He's been waiting for you the whole time," said Mr. Coleman. "He's good at charades, but he's terrible at hiding. Go on, get going."

CHAPTER 17

When Danny was twelve, he tried to impress a girl on his street by shinning to the top of a sycamore tree. The purpose of the mission, aside from showcasing his climbing skills—something he believed, like most boys his age, to be one of the defining characteristics of a superior boyfriend—was to rescue a cat that didn't even belong to the girl in question (he didn't know this at the time) and almost certainly didn't need rescuing (he had an inkling this might be the case, but he needed a pretext to demonstrate his pre-man manliness, and saving a cat from a tree seemed as good a reason as any). As if to emphasize this, the animal calmly waited for Danny to scale the most challenging parts of the tree until the two of them were well within spitting and hissing distance of one another, at which point the animal scarpered down the trunk and bolted up an adjacent tree while Danny was left teetering in the branches for just long enough to contemplate how stupid he looked before he lost his footing and tumbled to the ground.

His plunging body took a trajectory that miraculously avoided every branch, a divine stroke of luck that enabled him to limp away from the scene with his dignity shattered but his body intact. Danny often thought how fortunate he was to survive the accident at all, never mind with all of his appendages still working and his brain fluid still in his skull, and sometimes he had nightmares about that day, flinching awake in the darkest hours with the sense of falling still churning his guts. As with all good bad dreams about falling, however, he always woke prior to impact. But that night, when he crawled into bed after his session with Krystal and once again found himself tumbling towards the earth as the neighbor's daughter looked on in horror and the cat looked on in morbid amusement, not only did he hit the ground before waking, he also hit every branch on the way down. He lay there in agony at the bottom of the tree until the sound of his alarm delivered him from torment. When he groggily opened his eyes and tried to switch it off, the slightest movement hurt so much that he felt like somebody had broken into his apartment and beat him all night with a rolling pin. He wondered for a second how pain from a dream could migrate to the real world before his foggy brain caught up with him and he realized that his dream was in fact a manifestation of the actual pain that he now felt as a result of yesterday's visit to Fanny's.

Forcing himself onto his feet and into his slippers, Danny ignored Will's curious stares as he hobbled around the kitchen and made his breakfast before waving him off to school. Then, lowering himself onto the couch, slowly, as if he were entering a scalding-hot bath, he quietly took stock of the situation.

He wouldn't be dancing that day, that much he knew. Nor would he be dancing for the next few days unless he somehow developed Wolverine's gift of recovery. Still, given his laughable earnings so far, Danny felt confident that his temporary stasis would have no impact on his current financial situation. He did rue the wasted hours that he could have spent practicing, but unable to move so much as a finger without fearing it might fall off, he grudgingly accepted that whatever dance moves he wanted to try would have to be tried in his head.

Recalling his conversation with Krystal, he rolled off the couch and over to the television cabinet, where he searched through the various DVDs and computer games for Liz's copy of *Dirty Dancing*. She used to own it on VHS, but she'd watched the film so many times that the tape had worn out (particularly around the parts where Patrick Swayze appeared without his shirt on), so Danny had bought her the DVD for Christmas one year, although that didn't stop her from trying to wear out that copy as well.

He'd lost count of the number of times she'd asked him to watch it with her, seriously at first and then later jokingly when she realized it was never going to happen. Over time it became something of a running joke between them, with Danny responding with deliberately elaborate excuses whenever she suggested popping it into the DVD player. He'd always planned to give in one day, to surprise her when she least expected it by either agreeing when she asked him or perhaps even suggesting the idea himself. It had never occurred to him that he'd never get the chance, that a day would come when their running joke would no longer seem funny but instead unbearably cruel. As the opening bars of "Be My Baby" started playing over the title credits, all he could do was grab Liz's picture and prop her up on the couch beside him.

"Better late than never, right?" he said to his wife, blinking away the tears as Baby Houseman's opening monologue began. "No spoilers, okay?"

They sat there together for the next hundred minutes while Danny quietly took notes. He apologized every time he rewound a scene so he could scribble down observations about Jennifer Grey's footwork or Patrick Swayze's impossibly hairless body. He cheered when Johnny Castle rescued Baby from the corner. And when the end credits rolled, he clutched Liz to his chest and cried so hard that the neighbor's dog started howling. He cried for every time he'd stubbornly refused to dance with her, which was every time she'd asked, and he cried for letting her dance alone, something she actually quite enjoyed but still broke his heart when he pictured her now, in the middle of the dance floor, surrounded by strangers while he watched from the sidelines. He cried for caring more about making a fool of himself than he did about putting a smile on her face, and then he laughed, because he had to laugh, when he reminded himself that the man who wouldn't dance for fear of what people might think was now taking notes to improve his dancing-panda abilities. Danny had no idea how he had ended up here, but he knew that Liz would be proud of him. She'd be pissed that he'd waited so long to get his dancing shoes on, but she'd be proud of him.

Wiping his eyes with the back of his sleeve, Danny began to search for as many dancing films as he could find. He started with the ones that Krystal had mentioned and then made a list of all the others she hadn't. *Saturday Night Fever. Singin' in the Rain. Moulin Rouge! Step Up* (all of them—even the last one). *Silver Linings Playbook. Save the Last Dance. Magic Mike. La La Land.* And as many old Torvill and Dean clips as he could find. And that's all he did for the next two days, watching movies from the time Will went to school to the time he came home, and from the time Will went to bed until the time he woke up, rarely moving from the couch for any reason other than to use the bathroom, to forage some food from the kitchen, or to stretch his stiff but slowly healing muscles.

When he felt sufficiently recovered, Danny began to practice what Krystal had taught him as well as whatever moves he'd picked up from the films he'd recently marathoned through. He downloaded a metronome app and listened to it while doing anything and everything he could to synchronize himself with the rhythm. Brushing his teeth. Clipping his nails. Tapping his hand on the dinner table. Pacing from the front door to the kitchen and back. Nodding his head. Moving his shoulders. Nodding his head and moving his shoulders. Cleaning the windows for the first time since Liz had died, and then cleaning them again until the view didn't look so bad anymore. Scrubbing the stain on the carpet left by his dirty work boots. Chopping carrots into slices. Chopping the slices into pieces. Chopping the pieces into smaller pieces. They ate a lot of carrots during that time, something that neither Danny nor Will was particularly thrilled about, but whether it was down to the carrots or the metronome or Krystal or Johnny Castle or a combination of all the above, it was clear that something had changed when Danny returned to the park five days later. His dancing skills were still far from impressive, but they had at least evolved enough to draw a modest crowd. Also, unlike the handful of people who had stopped to watch him previously, many of whom regarded him in the same way they'd regard half a mouse lurking in their sandwich, this new wave of spectators actually seemed to enjoy his performances, if not tremendously then enough to part with whatever spare change they happened to be carrying. He made more money in that first day back at the park than he had in all the other days combined, and even though it was barely a fraction of what he owed Reg, it nevertheless felt good to receive some validation for his efforts.

As if his day couldn't get any better, his mood was further bolstered by the sight of El Magnifico storming towards him across the park in what appeared to be a purple dressing gown.

"Where is it!" he demanded. He looked pale and fragile without his robe, like a recently evicted hermit crab.

"Are you wearing a dressing gown?" said Danny, removing his mask and staring at the man's clothes.

"Yes, I am, and you know why, don't you, you furry little bastard."

"I hope you've got something on under there."

"Cut the shit, ferret, where is it?"

"Where's what?" said Danny, struggling to keep a straight face.

"You know what!" said El Magnifico. "Where's the robe?"

"What robe?"

"You know what robe!"

"I'm sorry, La Fantastico, but I have no idea what you're on about."

"You know that's not my name! And you know exactly what I'm talking about. You and Worzel fucking Gummidge over there took it." He pointed at Tim across the park.

"Somebody stole your robe? How terrible. You can't trust anybody these days, can you?" He pulled his mask down to hide his smile.

El Magnifico started to tremble as one eye began to twitch.

"Wait, are you just angry with me right now, or are you actually trying to set me on fire?" said Danny.

"You'll pay for this!" said El Magnifico, marching off with his dressing gown billowing behind him like an angry homeowner searching for the paperboy who'd mangled his *Telegraph*.

Danny sat down and chuckled to himself as he tied his shoelaces in preparation for his next performance.

"Hi," said a voice that was strange yet also somehow familiar.

Danny stared at his shoes, his skin prickling with excitement and his blood pumping loudly in his ears as he took a deep breath and looked up to find his son standing in front of him.

"Thanks again for the other day," said Will.

Danny nodded, unsure what else to do. An awkward silence descended.

"What are you supposed to be anyway?" said Will, coiling and uncoiling his tie around his hand. "A panda or something?"

Danny nodded again, vaguely aware that Will was the first person to correctly guess his species but too dazed to rejoice in the fact.

"Why don't you talk?"

Danny froze while he tried to figure out how to answer anything but a yes-or-no question. Noticing his bag by his feet, he took out the pad that he used to record his takings and scribbled something down in capital letters so Will wouldn't recognize his terrible cursive.

Because I'm a panda, read the message.

Will smiled. "I get it," he said. "You know, not wanting to talk. I don't talk either."

You sure about that? wrote Danny.

"Okay, I don't *usually* talk. You're the first person I've spoken to in over a year. And the first panda I've spoken to in, well, forever, I guess."

How does it feel?

"I don't know," said Will. He shrugged. "Normal. Weird. Both."

Why did you stop? wrote Danny, the anticipation of learning the answer to the one question that had plagued him since the accident causing his hand to shake slightly.

Will said nothing for so long that Danny started to think he'd lost him again.

"It's hard to explain," Will said at last.

Try, prompted Danny. *Pandas are great listeners.* He flicked his moth-ravaged ear for emphasis and accidentally ripped it off in the process.

"Another time maybe," said Will, returning Danny's rogue appendage. "I have to go."

Danny's pen hovered over the pad as he desperately tried to think of something to say that might keep the conversation alive, but Will was already halfway across the park by the time he'd finished writing. He looked at the paper and sighed.

Wait, read the note.

CHAPTER 18

"Hi, Will!" shouted Danny as he closed the front door and paced around the apartment excitedly. "Will? You home, mate?"

He'd left the park early that day, unable to focus on anything other than his unexpected encounter with his son. Even now, several hours later, the whole thing still seemed unreal, as if he'd just been magically cured of some terminal affliction.

He found Will playing with his iPad on his bed.

"There you are!" said Danny. He tried to look casual by leaning on the doorframe until he remembered that he never leaned on doorframes and was probably just making himself look suspicious. He stopped leaning. "How was your day?"

Will shrugged and returned to his iPad.

Danny continued undeterred. "Do anything interesting?"

Will shook his head without looking up.

"School okay?"

Will nodded, his eyes still fixed on the screen.

Danny changed tack. "What do you want for dinner?"

His son shrugged.

"We can have anything you want. Pizza. Burgers. KFC. Just say the word and it's yours."

Will put his iPad down and looked at Danny then.

"Anything at all," prompted Danny. He looked like a dog waiting for a tennis ball. "Just say the word."

Will's eyes narrowed. His mouth opened a fraction and for a brief moment he looked like he was about to say something. Danny leaned forward, keen to catch every syllable, but the only sound that emerged from Will was that which accompanied his powerful sneeze. He wiped his nose, shrugged, and returned to his iPad.

Danny slowly backed out of the room and closed the door behind him, quietly reprimanding himself for believing that things would be so simple. Still, he couldn't help but smile as he rummaged through the freezer that was long overdue for a defrosting and chipped a lasagna from the arctic wasteland. He was still smiling the next morning when he made his way to the park.

Danny danced for the biggest crowd he'd ever danced for that day. There must have been thirty people at least. None of them had gathered because his moves were slick, because they weren't. Nor had they gathered because his timing was impeccable, because it wasn't. What drew them over was the fact that even though Danny didn't have much to give, he still gave everything he had. He performed with an energy he didn't know he possessed, he moved with a confidence that far surpassed his abilities, and he danced without the ever-present fear that he looked completely ridiculous, which he did, but Danny didn't care, not that day. He didn't see the crowd that encircled him. Nor did he see El Magnifico glowering at him from across the park. He saw nothing, he heard nothing, and he felt nothing except for the music. When the track ended and Danny took a bow, it wasn't the sound of coins landing in his lunchbox that made him smile. It was the sound of applause—*real* applause—that warmed his spirit. He didn't even notice the money until the crowd had dispersed, and when he did he was shocked to find that his one performance of barely five minutes had brought him more than ten pounds in change.

By the end of the day he'd managed to make more than sixty pounds plus a handful of pennies, and by the end of the week he'd taken close to half of what he used to take home from the building site, and had a lot more fun doing it. For the first time since he'd become a dancing panda bear, Danny felt like what he was doing might not be so crazy after all.

"Blimey," said Tim when he saw Danny counting his money. "Maybe I need to start dressing Milton up as an animal."

"Milton *is* an animal," said Danny, nodding at Milton.

"People don't like cats anymore. They like pandas. You're the park's main attraction these days. Look at you, you're rolling in it!"

"Hardly. I still can't afford to pay my rent."

"Then maybe this might help." Tim pulled a flyer from his top pocket and handed it to Danny.

"What's this?" he said, staring at the piece of paper pinched between his furry fingers.

"Battle of the Street Performers. Four weeks' time. Hyde Park. First prize—"

"Ten grand!" yelled Danny. "Holy shit."

"That should help with the rent."

"Only if I win."

"So, make sure you win."

"Hold on," said Danny, checking his phone. "What time is it?"

"Close to four."

"Bollocks," said Danny, slapping the flyer with the back of his hand. "I'm too late. Registration closed at three."

"I know," said Tim. "Which is why I signed you up this morning."

"You're serious?" said Danny. Tim nodded. "I want to hug you right now. Can I hug you right now?"

"I wouldn't. Milton's got a bit of a jealous streak. He's basically the reason I don't have a girlfriend. Well, that and my face."

"Got it. Maybe just a handshake, then."

"Probably best," said Tim. The two men shook hands.

"Why are you helping me, by the way?" said Danny. "Won't we be competing against each other?"

"Yeah, we will, but more importantly, we'll be competing against El Magnifico, and the more competition he has, the less chance he has of walking away with the trophy. I don't mind losing to you. I just don't want to lose to David Tosserfield over there."

Danny stared at the flyer, sure that this was the closest he was ever going to get to the prize money. His odds of winning were slimmer than the rolled-up cigarette tucked behind Tim's ear, but he knew he had to try. He also knew that if he wanted to stand any chance of succeeding, he was going to need all the help he could get.

Music throbbed and strobe lights flickered as disenchanted women danced for loud men with slick hair and soggy collars. Danny edged his way through the crowd towards the bar where Vesuvius was busy serving customers. He looked at Danny and winked.

"Come for that kiss, have we?"

"I'm saving that for a rainy day," said Danny.

"That's not rain?" said Vesuvius, nodding at the wet patches all over Danny's shirt.

Danny looked down and grimaced when he saw he was covered in other people's sweat. "I wish," he said, grabbing a napkin from the dispenser and gently dabbing himself. "Is Krystal around?"

"Why do you think this place is so busy?"

The collective murmur of the crowd grew louder as the music began to fade out.

"You're just in time," he said, pointing over Danny's shoulder as the lights dimmed and the room went dark.

Danny turned to see several spotlights gathered on an empty podium with a pole in the middle and dark-red curtains behind it. People started shoving each other as they surged towards the stage, eager for a better view of the night's main attraction. The walls and floors began to shake with a bass line so penetrating that Danny felt almost violated by it, and moments later Krystal emerged, slinking through the curtains and strutting into the spotlights wearing a Stetson, cowboy boots, a holster hung from a belt on her hips, and black-and-white cowhide underwear that was so skimpy the cow probably didn't even know it was missing.

The men roared and whistled as Krystal approached the pole. One of them lunged for her leg as she passed and got a solid kick to the face for his efforts.

"That's my girl," said Vesuvius with fatherly pride.

Danny had only ever been to one "gentlemen's" club before. It was shortly after he'd started working with Alf, when one of the brickies invited everybody to his stag party. The event was supposed to be a pub crawl, and that's what it was until the sambuca started flowing and somebody suggested going to Sunset Boulevard, a notoriously dodgy lap-dancing club. After calling Liz and clearing the idea with her (he was hoping she wouldn't let him go, but she seemed to find the idea of Danny in a strip club hilarious), Danny reluctantly tagged along. He had no desire to stuff money he didn't have into thongs of women he didn't know, but he'd only been working on the site for a couple of weeks by then and he didn't want to be remembered as the only person who left the party early.

The entrance fee included a free lap dance, which Danny didn't want, so he donated his ticket to another member of the group, who in turn gave it to one of the girls and informed her that Danny was the rightful beneficiary. Before he realized what was happening, a bony teenager with straight blond hair and eyes that were darker than a Whitechapel alley planted herself in his lap. Not wanting to cause offense by asking her to unmount him, and physically unable to remove her without risking a pummeling from the bouncers who seemed almost eager for somebody to break the no-touching policy, he awkwardly endured the lifeless two-minute shuffle, his eyes fixed firmly on the ceiling while the girl jerked around on top of him, her hands on his shoulders and her gaze flitting between Danny and a grim-faced man who watched her from the corner. When it was over and he sheepishly tipped her, bypassing the garter she presented and handing her the money directly, the girl barely cracked a smile as she slid from his lap and climbed onto the next person the man in the corner jabbed his finger at. The whole experience was about as erotic as a trip to IKEA, which was where Danny would have rather been, even though he hated IKEA. As for the girl, she looked like she'd prefer to be bungee jumping without a rope than grinding away in the laps of strangers, but as Danny watched Krystal working the crowd, he couldn't help but notice how much fun she was having. She wasn't dancing for the men who were elbowing each other for the chance to cram her boots with their mortgage payments and children's tuition fees. She was dancing for herself, and they were paying her to do it.

"Beautiful, isn't it?" said Vesuvius. Danny knew precisely what he meant. Krystal wasn't just performing. She was psychoanalyzing, scanning faces, observing body language, profiling personalities, identifying weaknesses, searching for voids that needed filling, pressing buttons that were rusty with neglect, playing people against one another. She could trick the room's most frugal man into reaching for his wallet by simply paying more attention to the man standing next to him. She could pluck the last tenner from a poor man's hand and still somehow manage to make him feel rich. She could make the biggest loser in life feel like he'd won the lottery with nothing more than a well-timed wink, and when she saw Danny across the room and flashed him a fleeting smile, even he experienced a warm fuzzy flutter that lasted until she finished her performance and joined him at the bar.

"I thought you promised to leave me alone," said Krystal. She took a thirsty gulp of the water that Vesuvius handed to her.

"I have something I think you might be interested in," said Danny at the same moment that Fanny walked past. She shot him a dubious glance. "And, no, Fanny, it's not what you think." She smirked and disappeared into the cellar. "It's this." He unfolded the flyer and slapped it onto the bar.

"What am I looking at?" said Krystal, staring blankly at the piece of paper.

"Battle of the Street Performers. Winner gets ten grand."

"Thanks for stating the obvious, Danny. I mean, why are you showing it to me?"

"Because I'm going to enter," said Danny.

"Good for you."

"And I'm going to win."

"That's the spirit."

"Because you're going to help me."

"Oh yeah?" she said. "By doing what exactly? Murdering the other contestants?"

"By teaching me everything you know."

"In four weeks?"

"Yeah. Well, three and a half."

"Have you been drinking?" said Krystal. She turned to Vesuvius. "Has he been drinking?" Vesuvius shrugged.

"I'm serious," said Danny. "I think we can do it."

"No, Danny, *we* can't."

"We can split the winnings. Fifty-fifty. Straight down the middle."

"It's not possible."

"Yes, it is. You just divide ten by two, it's super easy."

"No, you muppet. I mean there's not enough time."

"Okay, how about sixty-forty?" he said.

"Danny, it's not about the money, it's—"

"Fine, seventy-thirty, but that's my final offer."

"You still owe me a hundred quid!" she said.

"Plus the hundred quid, obviously."

"Danny, I'd help you out if I could, really, but you're not even close to competition level and there's no way in hell you're going to get there in the next few weeks, even if we practiced twenty-four hours a day. If I was you, I'd forget about the contest and focus on perfecting the simple stuff I taught you." She threw her empty bottle in the bin and adjusted her cowhide bra. "I need to get back to work."

"Wait!" said Danny as Krystal turned to leave.

"Sorry, Danny."

"Look, just hear me out for one second. Please." Krystal sighed and gestured for him to finish. "I have a son. His name is Will. He's eleven years old. He was in the car with my wife when she died and he hasn't spoken a word since. Literally, nothing, so it's fair to say he doesn't need any more problems in his life, which is why I haven't told him that I lost my job, and I certainly haven't told him that I'm now a full-time fucking panda bear. He still thinks I work on the building site and he still thinks we can pay the rent, but we can't, and in four weeks' time my nasty bastard of a landlord is going to evict us, but not before he breaks whatever part of my body he deems to be my favorite, because that's the kind of nasty bastard he is. And the only way I can stop that from happening is if I win this competition. I know it's a long shot, and I know it's almost certainly hopeless, but I've got to try, because if I don't, I am well and truly fucked. So, please, help me. I've never begged for anything, but I'm begging you right now."

Krystal shook her head, but it wasn't a no. It was a "what furry little animal did I kill in my former life, or what frail old grandmother did I defraud to get punished like this?" kind of shake. She looked at Vesuvius, who'd been listening in on the whole thing.

"Well?" she said.

Vesuvius looked at Danny. Danny did his best to look pathetic, something he was becoming increasingly adept at. Vesuvius looked at Krystal and nodded.

"Seriously!" she said, her hands held out like Krystal the Redeemer. "I expected more from you, Suvi." She sighed and looked at Danny. "Okay, fine, whatever. I'll help. Just as long as you know that we're not going to win."

"You mean it?"

"That we aren't going to win? Absolutely."

"That you'll help me," said Danny.

"I said so, didn't I? Monday. Eight a.m. Don't be late."

"Thank you," said Danny. "Seriously. You don't know how much this means to me. You are such a . . . WANKER!"

"What the—"

"Not you!" he said before Krystal could hurt him. "That guy!" He pointed to a man with a bony face and a deathly pallor who was standing behind Krystal. "That one. Right there. That's the guy who got me fired."

Three men in black suits were talking together near the stage. They looked like they'd just come from a funeral, and Viktor looked like the deceased, his already pale complexion almost translucent beneath the cold white light he was standing under.

"So this is all *his* fault?"

Danny nodded, his jaw flexing as he stared at Viktor the way El Magnifico stared at something he wanted to ignite.

"Suvi," said Krystal, "pass me that mic, would you?"

She took the microphone from Vesuvius and shoved her way through the door behind the bar. Danny was still wondering where

she'd gone when everyone in the room started cheering and chant-
ing. He turned to see what the fuss was about and found Krystal
standing in the middle of the stage, surrounded by a sea of punters
who wrongly assumed they were getting an encore.

"You all having a good night?" she said, aiming the microphone
at the crowd.

Everybody roared in agreement.

"I thought you might say that. Anybody up for a little game?"

Another boisterous chorus of approval.

"The prize is one private dance with yours truly . . ." she said.
She waited for the room to calm down before continuing. ". . . and
the winner is the first person to unblock the men's toilet using that
guy's head."

Vesuvius cut the lights and turned a single spotlight on Viktor.
The man raised his hand to shield his eyes from the glare, but to
everyone else it looked like he was inexplicably trying to identify
himself, as if being used as a human toilet brush was something
he actually quite enjoyed.

"May the best man win!" she shouted as everybody piled on
Viktor, who turned an extra shade of pale as the crowd dragged
him off to the bathroom.

CHAPTER 19

A man with what appeared to be a single giant dreadlock slowly pushed his rickety drinks cart along the path. Danny watched him from his place on the bench, his soggy panda mask steaming gently beside him after a flustered and ultimately misguided attempt to entertain the crowd with his own rendition of the Gangnam dance.

He bought a can of Pepsi (which on closer inspection turned out to be something called Popsi) and held it against his forehead while he watched an elderly lady in a cardigan, far too thick for such a sunny day, trying to attach a leash to a hyperactive beagle. Every time she came within arm's reach of the animal, the beagle would gallop off and patiently wait for her to catch up before repeating the process ad infinitum. Noticing Danny, either by sight or by the curious odor that continued to emanate from his costume no matter how many times he washed it, the dog trotted over to investigate, sniffing his furry leg as if it wasn't sure whether to bite it, hump it, or use it as a pee-post. It was still trying to decide which course of action to take when the old lady seized her moment to creep up from behind and fumble the leash onto the distracted dog's collar. She gave Danny a knowing nod, as if the two of them had been catching dogs together for years. Danny nodded back and watched the lady shuffle off while her beagle kept trying to trip her up with the leash.

He'd barely put his mask back on when somebody spoke behind him.

"Do you just sit in the park all day?"

Danny fumbled for his pad and pen with his cumbersome panda paws as Will appeared in front of him.

It's nicer than sitting in the middle of the motorway, he wrote.

"No, I mean, don't you have a job or something?"

I'm a panda. This is my job.

Will smiled. He took off his schoolbag and sat on the bench.

Don't you have a job? wrote Danny.

"Yeah," said Will, removing his tie and wrapping it around his hand. "School. I work long hours and don't get paid. It's the worst job ever."

I think I prefer my job.

"I'd prefer your job too," said Will. "Except today." He squinted at the sun. "It's too hot to be a panda today."

It's okay. Pandas have complex cooling mechanisms.

"Oh yeah? Like what?"

Danny held up his can of Popsi.

Will rolled his eyes. "Very complex," he said.

Danny stared at the unopened can and wished he could take a sip without blowing his cover.

"Mr. Carter's Complex Conundrum," said Will, as if to himself.

Danny looked at him, confused.

"It's this thing in my maths class. The teacher, Mr. Carter, he always writes a problem on the board at the start of every lesson, and at the end he picks somebody in the class to solve it. He calls it his Complex Conundrum. I hate it."

Why? wrote Danny, unsure where this was going exactly but eager to keep Will talking.

"Because I can never figure it out. Sometimes I know the answer, but most of the time I don't, so whenever it's something I don't understand, I stay quiet and put my head down and hope he doesn't see me." Will thought for a minute. "It's hard to explain, but that's sort of why I stopped talking."

Because of maths class? wrote Danny. He made a mental note to find Mr. Carter and give him a complex conundrum of his own.

"No," he said. "Not because of maths class. Because, well, something bad happened last year. Something *really* bad, and it just didn't make any sense." He pulled on the end of his tie and it closed around his hand like a mini boa constrictor. "It was like Mr. Carter's maths problems but a million times worse. I didn't know what to do, so I just did what I always do in maths."

You stayed quiet and hoped people would leave you alone? wrote Danny. Will nodded. Danny scrapped his mental note to wait for Mr. Carter in the school car park.

"I just thought it would all go away if I ignored it for long enough. Like, as long as I didn't attract any attention to myself, then the problem would just, I don't know, disappear or something." Will unwound his tie and started wrapping his hand up again. "It seemed a lot more normal when I said it in my head. Now it just sounds weird."

And there it was. Just like that. After fourteen months of wondering, Danny had an answer. He sat back and waited for the relief to wash over him, but when it came it was more like a drizzle than the downpour he expected. It was sadness he felt more than anything—sadness that Will had been suffering in silence, sadness that Danny himself had let it happen, and sadness that it took this ridiculous situation to finally learn the truth.

Realizing that Will was staring at him, he grabbed his pad.

It's not weird, he scribbled.

"Thanks."

Talking to a panda is weird.

"My teacher talks to a one-eared rabbit called Colin." Danny didn't even bother trying to respond to that one. "And anyway, if talking to a panda makes me weird, what does that make you?"

Super weird.

"Then I guess we're just two weirdos in a park," said Will.

I'm okay with that.

Will smiled. "Me too," he said. Danny reminded himself to talk to Will about the dangers of hanging out with weirdos in parks.

He looked at his son, trying to gauge his next move. The polite thing to do, under normal circumstances, would be to ask about the incident last year, but Danny was already painfully aware of that part of the story and he didn't want to force Will to talk about anything he didn't want to (although if there was one thing the last year had taught him, it was that nobody could make Will talk about anything he didn't want to). Then again, perhaps he *did* want to talk about it, and by not asking, Danny was denying his son the first and maybe only opportunity to open up about the whole thing. Danny stared at the pad in his hands, unsure what to do next.

"Has anybody you know ever died?" said Will.

Danny wondered how best to respond. He didn't want to blow his cover by telling Will the truth, but he didn't want to lie to him either. He felt bad enough about the situation already.

Yes, he wrote, hoping Will wouldn't probe any further.

"Do you miss them?"

Danny nodded. *Very much.* He listened to a pair of sparrows chatting away in the branches above while his son quietly fidgeted beside him.

"My mum died in a car crash," said Will. "It was over a year ago, but I still miss her a lot." His voice sounded distant, as if trying to retreat. "It feels strange, saying that out loud."

Danny's pen hovered over the pad. There were plenty of things he could say, things that people had mumbled awkwardly to him at the funeral and intermittently ever since ("Sorry.... I'm so sorry.... I'm sorry for your loss.... Heaven needed another angel"—Christ, how he hated that one); but as much as he knew that people meant well, he also knew that none of those words made the slightest bit of difference to how he was feeling.

I bet she was an awesome mum, he wrote.

"She was," said Will. He pulled a thread from his sleeve and let it fall to the ground. "She was the best mum ever."

They sat in silence for a moment, staring in different directions but both focused on the same person.

"It's rubbish when people die, isn't it?" said Will.

Totally rubbish.

Danny watched the man with the drinks cart trying to fix his broken parasol. Every time he straightened the ribs on one side, the other side would collapse, and just when he'd managed to get them all open, a mild gust of wind knocked them all down again. A passing couple laughed, but Danny knew how hard it was to hold things up when everything felt like it was falling down.

"You want to hear something stupid?" said Will. Danny gestured for him to continue. "I want to talk to her about what happened. I just want to talk to my mum about my mum not being here any-more. She was always the person who could make this kind of stuff better."

Danny nodded. He wiped his eyes with the back of his hand, forgetting he still had the mask on.

Can't you talk to your dad? he wrote.

Will shook his head. "It's not the same. Mum was my mum, but she was also my friend, you know? But Dad, well, he's just my dad. He was always working a lot, going out really early and coming home really late, so it was often just me and Mum who did stuff together."

What sort of stuff? wrote Danny.

Will shrugged. "Just hanging out. We went to Brighton once. That was cool. And Stonehenge. And she used to make these amazing pancakes. She had this secret recipe that her gran gave her. She keeps it hidden in a cookbook. Sometimes I look at it, I don't know why. Just to check it's still there I guess. When she died it was just me and my dad, and it felt, I don't know, different."

Like you were living with a stranger?

"Yeah. Like strangers. He doesn't really know anything about me. He still thinks I like Thomas the Tank Engine, even though I'm in secondary school. He thinks I like peanut butter, even though I hate it. And we never do anything together, not like I used to do with Mum. He hardly even *talks* about her."

Maybe he finds it too painful, wrote Danny, wishing he could cross out the *maybe.*

"Maybe," said Will. "Or maybe he just wants to forget about her."

Danny's pen scratched loudly against the pad as he frantically scribbled something down.

That's not true! read the message. Danny had underlined it twice.

Will frowned. "How do you know?" he said.

Danny wanted to grab Will by the shoulders and explain to him that he hadn't forgotten Liz, that he *couldn't* forget her, ever, and that even if he lived for a million more years, even if he lived until the world started to fracture and break apart piece by piece until everything that remained had been scattered to the farthest reaches of space, even then she'd still be with him, keeping him company as he wallowed in the great unknown, happy to face the infinite darkness just as long as she was by his side. But he knew he couldn't say any of that, so instead he wrote the first thing that came to mind.

Because I'm a panda.

Will smiled. "Whatever," he said, standing and hooking his schoolbag over his shoulders. "I should probably get moving."

Good talking to you.

"You didn't say anything."

Danny scribbled out the word *talking* and replaced it with the word *listening*.

"Better," said Will. "See you."

He made his way across the park, his blond hair dancing in the breeze. Danny watched him go, his small frame growing smaller and smaller until he disappeared from view. Only then did he take off his mask and bury his face in his hands.

CHAPTER 20

Later that evening, once Will had gone to sleep, or pretended to have gone to sleep while secretly playing on his iPad, Danny flicked through the notebook that he'd used to scribble down his panda messages. In between his own words he'd hurriedly written as much of Will's half of the conversation as he could remember so that he was left with a document that, although far from perfect, nevertheless served as some kind of record of what could be the last conversation he might ever have with his son. Certain words and sentences were underlined roughly. *"He doesn't know anything about me." "Strangers." "Maybe he wants to forget about her." "He never talks about her anymore."* Each one hurt him in different ways, but the one that inflicted the most pain was the single word he'd encircled several times. *"Mum was like my friend, but my dad is just my dad,"* read the full sentence, but it was that one word—*"just"*—that Danny kept returning to. He didn't find it hurtful because it was cruel or unfair. What hurt him was that he knew it was true. He *was* just his dad. He wasn't a friend. He didn't know him like his mother did. Liz was like an encyclopedia of Will. She knew everything from his shoe size to his opinions on who would win in a fight between a stegosaurus and a triceratops. She knew the way he liked his hair cut. She knew where he was most ticklish. She knew the names of every single one of his stuffed toys from when he was younger, even when Will himself had forgotten them (or pretended he'd forgotten them). She knew his favorite foods, she knew his favorite color, she knew what he was afraid of, and she knew what she was most likely to find in his pockets on any given day. She knew which chocolate he would choose if presented with a tin of Cadbury Roses and she knew what he'd do with the wrapper afterwards. If Will disappeared in a time machine, Liz knew which period in history she'd most likely find him in, as well as which castle. She knew what dessert he'd choose in a restaurant, she knew what sort of surgery he'd perform on a Burger King Whopper, and she knew what piece he would choose in Monopoly as well as what streets he would buy. But having spent so much of Will's life working all the hours he could get, not because of his love for the job but because he wanted his son to have a better life than his own, Danny knew none of these things. It had never occurred to him that by working all the time he was in some ways denying Will a father. Not that there was much he could have done about it. Overtime wasn't a choice for Danny. He couldn't pick it up when he needed it and pass it up when he didn't. Danny *always* needed it, so he always took it. Liz sometimes joked that he was a workaholic, and he was, but they both knew that his motivation had nothing to do with greed and everything to do with necessity. Even with Liz's income, their combined salaries left little in the way of savings once their rent and their bills and their groceries had been taken care of, and things had become even tighter since Liz had passed away, which was why he'd been working more than ever.

Money wasn't the only reason, though. Danny had once worked with a joiner whose daughter had drowned on holiday. One minute she was playing near the shore and the next minute she was gone, dragged out to sea by a vicious current that only let her go when it was far too late. The man returned to his job just two days afterwards, something that Danny struggled to comprehend. He thought that work would be the very last thing on anybody's mind in the wake of such an unspeakable tragedy, but when he lost Liz, he finally understood why the joiner had done what he did. In a time when nothing made sense anymore, a time when your mind stopped being your friend and became your worst enemy, sometimes work was the only thing that stood between you and insanity. Being on-site let him switch off his brain. He could hide his mental remote in his locker and leave it there for the better part of the day. Working allowed him to forget, if only for a little while and only until the night arrived and brought with it a darkness that he struggled to shake off even after the sun had risen. Just as Will had chosen silence, Danny had chosen work. They'd been coping, separately, in their own ways and in their own time, for the last fourteen months. That's what Danny had assumed they'd been doing, at least, although whether that was because he genuinely believed it or because it simply allowed him to wallow in his own self-pity he really couldn't say, but after talking to Will, he realized that his son *hadn't* been coping at all. Silence wasn't his way of dealing with things. If anything, his silence was something he chose in the *absence* of a coping strategy.

It was then that Danny understood how Liz's death had left not one void in their lives but two. There was the gaping hole she'd left inside their family, but there was also the hole she'd left *between* them, a hole that Will had filled with silence and Danny had filled with work when they should have been filling it with each other. Liz had in many ways been the bridge that connected the two of them, and they'd been living on different sides of the same ravine since the day that bridge had collapsed, watching each other from a distance while the space between them grew ever wider. Soon it would be so wide that they'd lose each other forever if Danny didn't find a way to close that gap, and quickly.

Gripped by a belated sense of urgency, Danny trawled through his notes and reread everything that Will had told him in the park. He'd already lost so much time that he didn't want to wait another second to implement some type of positive change in their relationship, but while everything that Will talked about was doable, nothing was doable *now*, at 10 p.m. on a Tuesday evening. He couldn't exactly drag Will out of bed and take him on a surprise visit to Stonehenge. Nor was it an ideal time to take an impromptu trip to the seaside.

Suddenly he had an idea. He went into the kitchen, opened the cupboard, took out the jar of peanut butter, and threw it in the bin.

"A journey of a thousand miles," he said to himself. He couldn't remember the rest of the saying but he was pretty sure it was relevant here.

Noticing that a bag of flour had also expired by a considerable margin, he threw that in the bin as well. As he did so, he remembered Will talking about the pancakes that Liz used to make. She'd always maintained that the recipe was a secret passed down by her grandma, which had been passed down by *her* grandma and so on and so forth, back to the very dawn of creation, but the recipe wasn't really a secret and it hadn't been given to her by her grandma. She'd soaked a piece of paper in a tray of tea overnight and dried it to look like parchment. On it she'd written a Jamie Oliver recipe she'd copied verbatim from his website and written *Gran's Top Secret Pancake Recipe* across the top, along with several faux warnings akin to those on treasure maps informing the reader of all the curses and plagues they'd incur if they dared sneak a peek at the information within.

Danny studied the shelf above the microwave, where several cookbooks were gathering dust. He slowly leafed through them, pausing occasionally to read the funny little notes-to-self that Liz had left at the top of several pages (bean casserole: *Hearty but farty*;

Bolognese: *Save some wine for the dish next time*; homemade pasta: *Only for sadists*; homemade ketchup: *Why did I even bother?*), until he found the recipe he was looking for.

Making a note of the ingredients—mostly just standard pancake ingredients, but Danny knew nothing about cooking so he copied them all word for word—he closed the book and returned it to the shelf. Then, sneaking out to the corner shop, he returned with as many eggs and as much flour and milk as he could carry.

Standing on a chair and immobilizing the smoke alarm in anticipation of the disaster he was sure was about to unfold, he pulled on Liz's apron, tied it at the back, floured his hands, wiped off the flour when he realized he didn't need to flour his hands, and started to cook.

Danny made a lot of pancakes that night. The first batch kept burning, the second batch refused to cook, and the third batch kept sticking to the pan. When he figured out how to stop that from happening (more butter), the next pancake stuck to the ceiling. He made close to twenty before he managed to get one onto a plate, but his satisfaction lasted only as long as it took him to taste it and realize he'd added too much salt. He tried again (and burned them again), and then he tried again (and undercooked them again), and then he flipped a couple more onto the floor, until finally, at around 2 a.m., Danny crawled into bed, burned, bruised, and quite literally battered, but confident he knew how to make a decent pancake.

When Will woke up the following morning and shuffled past the kitchen, he paused, frowned, and sniffed the air a couple of times. Shuffling backwards like a moonwalking zombie, he came to a halt in the doorway.

"Morning, mate," said Danny over his shoulder, his body blocking the hob so Will couldn't see what he was doing. "Sleep okay?"

Will didn't respond, too overwhelmed by the sounds and smells that were emanating from the kitchen.

Danny smiled. "Take a seat," he said. *"Le petit déjeuner est prêt."*

Will stared blankly at Danny.

"It means breakfast is ready. It's French."

A single nod, but still no movement.

"Look, just go and sit down, okay?"

Will's confusion only increased when he saw the maple syrup. He knew what it was, and he knew what it was for, but he couldn't figure out what it was doing on their table. He was still staring at the bottle when Danny emerged from the kitchen with a teetering platter of pancakes. Will looked at the mountain of food, as speechless as a boy who didn't speak could be.

"What?" said Danny as he put the pancakes down with a thud.

Will looked at the table and turned his palms up.

"I thought it was time for a change," said Danny. "What do you think?"

Will nodded as if Danny had asked if he wanted a raise in his pocket money.

"Great," said Danny, disappearing into the kitchen. "One second, let me get the plates."

He grabbed Will's Thomas the Tank Engine plate and mug from the drying rack and looked at them in turn. Then, holding them at arm's length and opening his hands, he dropped them both and watched them shatter against the kitchen floor.

"It was an accident," he said when Will ran in to see what the commotion was about. "My hands were wet and . . . they just slipped. I'm so sorry, mate."

Will picked up the mug and slowly turned it around in his hand. It was still intact except for the handle that had broken off and skidded beneath the cooker.

"I'll buy you another, I promise," said Danny, who couldn't quite gauge Will's expression and thought that perhaps he'd terribly misjudged everything and destroyed something that was in fact very dear to his son. Any concerns he might have had were quickly extinguished, however, when Will raised the mug above his head and slammed it down on the floor. This time James the Red Engine did not survive.

"Or . . . not," said Danny as Will grinned at him. "Come on, help me get some plates before the pancakes get cold."

Krystal was doing side splits in the middle of the floor when Danny arrived at the dance studio.

"Please tell me that's not part of the lesson," he said, wincing on her behalf.

"You said you wanted to learn everything," said Krystal without looking up.

"I meant everything that wouldn't require physiotherapy afterwards. Or psychotherapy, for that matter."

"Don't worry," she said, bending forward until her body was flat against the floor. "We're not doing the splits." She waited for Danny to breathe a sigh of relief before adding: "Splits are for the next session." Danny laughed. Krystal didn't.

"You're joking, right?"

"I guess you'll find out tomorrow."

"If I survive until tomorrow. I've barely recovered from last week's lesson," he said.

"That was nothing, mate. The last lesson was like a walk in the park with your grandmother."

"I hated my grandmother."

"Oh. Well, *this* lesson is going to be like a walk in the park with your grandmother, then."

"Are you going to hit me with a stick and tell me what a hideous mistake I was?"

"It can definitely be arranged," she said, pulling her legs in and springing to her feet. "Come on, you muppet, get stretching. We ain't got all day."

Danny peeled his coat off and chucked it into the corner.

"I'm never going to win this competition, am I?" he said, wobbling about on one foot while he stretched his hamstring.

"Nope."

"Can't you just lie to me for once?"

"I'm not a good liar."

"Try."

"Okay. Fine." Krystal thought for a moment. "You're very attractive."

"That's not what I meant."

"What do you want me to say, Danny? You know as well as I do that this is a long shot. Longer than a long shot. The bloke from *American Sniper* would have a hard time pulling off this kind of long shot, and he was a fucking sniper. So if you're asking me if I think you're going to win this competition, then no, I don't. But if you're asking me if I think you have a *chance* of winning, then yes, I do. Not a big chance. Barely a hair on the arse of a chance. But a chance nonetheless. I wouldn't be here if I didn't, Danny, because despite how funny it is to watch you stagger around this studio like a drunken donkey, I'd much rather be spending some quality time with my Netflix subscription than stuck here with you. And anyway, it's not like you're training for *Swan Lake* or anything. You don't need to learn the entire fucking history of dance. You don't need to be the next Michael fucking Flatley. You just need to know enough to get through one single performance, which will only last a few minutes max, and don't get me wrong, that still equates to a shit-barrel of work between now and then, but you've got three whole weeks to nail a three-minute dance routine, which is one minute per week, which is, like, what? Ten seconds per day or something, so if you follow my lead and do as you're told and stretch your sodding hamstrings properly"—Danny instantly improved his posture—"then who knows, we might just be able to give those fuckers a run for their money. There, how did that sound?"

"Surprisingly uplifting," said Danny.

"Then I guess I'm better at lying than I thought," said Krystal. She gave him a wink. "Anyway, watch and learn. I've got a routine that I think might do the trick."

The walls began to shake as Krystal pressed play on the stereo. She walked into the middle of the room and stood facing the mirror, rocking her head from side to side and shaking her arms like a boxer readying to rumble. The music grew louder as it built towards a crescendo, and when the beat eventually dropped, so too did all of Danny's expectations of ever getting his hands on that prize money. He watched Krystal flit across the floorboards with such agility that the wood barely creaked because her feet barely touched it, and when they did they fell like rain, somehow landing in several places almost simultaneously. She moved with the grace of a figure skater and the confidence of a free climber, the attitude of a rock star and the freedom of someone who followed nothing except the music. As the beat sped up towards a second crescendo she began to dance so fast that the lights started to flicker, perhaps because Fanny had forgotten to pay the electricity bill or perhaps, as Danny suspected, because Krystal's impossibly vigorous performance was somehow draining all the power from the neighborhood. Even when the music had finished and Krystal was no longer moving, Danny had the strange sensation that the room was still in motion, as if her body had generated a surplus of energy that was now spinning wildly around the studio in search of a way out.

"So?" she said. "What do you think?" She'd barely broken a sweat, but Danny's forehead was glistening.

"I think you're right. I'm never going to win this competition."

"You didn't like it?"

"No, I didn't like it," he said. "I *loved* it. You were incredible. But that's just the problem. There's no way I can learn all of that in three weeks."

"Not with that attitude, you won't," she said.

"Can't we do something easier?"

"You can do whatever you want, Danny. You can go out on that stage and fart 'The Blue Danube' through a rolled-up newspaper for all I care. It's not me that's going to get smashed into cat litter if you lose."

Danny stared at his reflection and tried to see anything but a panda facing extinction.

"You really think I can do this?" he said.

"We'll never know until you stop bitching and start dancing, will we?"

So that's what they did for the next three hours, and that's what they did every day that week. They worked from eight till eleven, after which time Krystal had to start work and Danny had to go to the park where he'd dance until the late afternoon before returning home exhausted. Even then, after Will had gone to bed, he'd quietly go over that day's lesson, tiptoeing through the various moves that Krystal had taught him in the kitchen while he tried not to crash into anything. He downloaded the song she'd picked, an aggressive piece of electronica that sounded like it came from *Guantánamo's Greatest Hits*. He added the track to his iPod and reluctantly listened to it wherever he went, tapping out the rhythm on his knees, the table, the seat of the bus, and any other surface he could find until he started to learn the structure by heart. Danny also included the song on his park playlist so he could practice while working, but the music seemed to repel a number of demographics (the middle-aged, the elderly, the people held without charge for terrorism-related offenses) and attract certain others that he didn't particularly want to attract, like the gang of teenagers who forced him to pose for several humiliating selfies before reimbursing him with his own money that they'd taken as collateral. He would have let them keep it had it happened a few weeks previously; but unlike then, when his lunchbox seemed to be nothing more than a storage area for beer caps, buttons, chocolate limes, and pocket fluff, Danny was making more money than ever. Even though he was still several hundred pounds short of what he needed to pay Reg, he tried not to think about that and instead did his best to clear his mind of everything except for the upcoming contest.

CHAPTER 21

Saggy sun worshippers sizzled in deck chairs while children played chicken with the waves or poked around in crab holes. Seagulls pecked at Styrofoam trays stained red and yellow with ketchup and curry sauce. Loutish youths swigged beer from cans and jeered at people beneath the flapping shadows of the flags that flew from the bright white turrets of Brighton Pier. Halfway along the promenade stood a noisy games arcade where Danny and Will were busy beating the hell out of each other, their eyes darting left and right as they followed the two fighters who were leaping about the screen in front of them. Will looked focused. Danny looked perplexed. Both were hammering the buttons on the console, but only Will seemed to know which ones he was hammering and why.

"Okay," said Danny. "Just. Hold on. If I. Wait a. There! Ha! Kick! Again. Again!"

He jerked the joystick left and right and mashed the buttons with his fist. His fighter threw out a combo of moves that looked impressive but achieved nothing. Will retaliated with a flurry of violence that left Danny's fighter bloodied and immobilized. A sinister voice-over implored him to finish the job.

"Is that it?" said Danny. "It's over? Again?"

Will nodded. Danny sighed.

"Go on, then, put me out of my misery."

Will hunched over the console and grinned like a deer with a trucker in his headlights as he rapidly wiggled the joystick and jabbed several buttons in sequence. His fighter hit Danny with a brutal uppercut that took his head clean off his shoulders and sent it bouncing down a ravine.

"Happy now?"

Will nodded, clearly proud of himself.

"Come on, I think I've been decapitated enough for one day."

Getting to Brighton hadn't been easy. The journey itself had been simple enough but the planning had been somewhat trickier. Ever since Will had told the panda about Liz taking him to the beach, Danny had been trying to figure out how he could propose a trip without Will getting suspicious. He'd already become an overnight pancake guru and Thomas the Tank Engine assassin, and he was afraid that Will would connect the dots if he suddenly suggested a trip to Brighton. In the end, it was Mo who saved the day. His parents had dragged him to Clacton-on-Sea to visit relatives for the weekend and Will was left with nothing to do, so Danny had proposed their own trip to the seaside and Will had eagerly accepted.

They weaved their way through the various machines in search of another game. Danny kept his eye out for one in particular, a ramshackle penny pusher that was already an antique the last time he'd seen it almost thirteen years ago. He knew it was probably long gone by now, broken down for scrap or languishing on eBay because nobody could afford to pay the shipping on a machine that weighed the same as a pregnant cow—but there it was, hiding in the corner, repelling the latest of a lifetime of assaults as two young boys tried to make it spill with a series of hip shunts and body checks.

Danny smiled when he saw it, recalling his first official date with Liz when the two of them jumped on the train without paying and spent the journey to Brighton locked in the toilet while Liz pretended to be an old lady suffering from alternating bouts of food poisoning and motion sickness whenever the conductor pounded on the door. They spent the day shoplifting from joke shops, being expelled from liquor stores, throwing chips near unsuspecting people so seagulls would attack them, and roaming the very same games arcade that he and Will were now roaming. Danny had also attempted to defeat the penny pusher via underhand means, not to get the money itself (that was a secondary objective), but to demonstrate to Liz how strong he was. All he succeeded in demonstrating, however, was that fifty kilos of skin and bones were no match for a half-ton machine, although he did at least manage to trigger the alarm, something he felt perversely proud of until he saw the security guard lumbering towards him. Bolting for the exit and realizing Liz wasn't with him, Danny ran back and grabbed her before they burst through the doors and escaped together down the pier. That was the first time they ever held hands, and they didn't let go for the rest of the day.

He traced his palm with his fingertips as he tried to imagine her hand in his, but before the sensation returned to him, Will thumped him in the arm and pointed at something nearby. Danny followed his finger and saw that Will was challenging him to a dancing game. He smiled.

"You're on."

They both stood on the dance mat and faced the screen, where two characters with multicolored hair and oversize features were waiting to guide them through the game.

"You know," said Danny, "I think it's only fair to tell you that I'm something of a master on the dance floor, so if you want to chicken out, now's your chance."

Will looked at him the way a bouncer looks at a drunk person who's trying to convince them they're sober despite not wearing any trousers.

"Suit yourself," said Danny, grinning like he'd already won. "Don't say I didn't warn you."

The game commenced and they started to dance, moving in time with the characters on the screen. The moves were basic at first, the types of maneuvers that Danny would have struggled to pull off just a couple of weeks ago yet now found tediously simple, but they soon became more complex as the characters danced with ever-increasing speed. He missed a few steps and dropped a few points but shrugged it off, certain he was still on course to reap his revenge for all of those earlier beheadings, until he looked at the other side of the screen and saw that his son hadn't missed a single beat. Danny redoubled his efforts, but the harder he tried, the more mistakes he made and the more points he lost, while Will's scoreboard continued to skyrocket. The longer the game dragged on, the more uncoordinated Danny became as he stopped paying attention to his own digital dance coach and focused instead on Will's flawless footwork. A crowd formed around them as accolades started flashing up, all of them for Will (*Masterful! Genius! Godlike!*), who didn't seem to notice that Danny had already conceded defeat. Nor did he notice the crowd around him until the game finally came to an end and everybody cheered, Danny loudest of all as he proudly watched his son, amazed by his skills and stunned by the contrast between the boy he thought he knew and the confident, lively, happy boy in front of him.

Danny grimaced as he watched an old man in the corner of the café lick not just his fingertip but everything up to the second knuckle every time he turned the page of his newspaper. Another, older man dozed upright beside his wife, who was delicately unwrapping a mint as if she were defusing a bomb. Above them fishing nets dangled from the ceiling like giant cobwebs, and actual cobwebs hung from the fishing nets as well as from the life rings, wooden boat wheels, and other nautical-themed décor. The place looked more like a boat ready for the breaker's yard than the Jolly Cabin it purported to be.

"You've played that dancing game before?" asked Danny across the table.

Will shook his head and took a slurp of his Coke.

"Then where did you learn to dance like that?"

Will shrugged. He fished an ice cube from the glass with a spoon and popped it into his mouth.

"Well, you're a natural, mate. You must get it from your mum. You certainly don't get it from me anyway."

Will smiled and crunched the ice cube between his teeth.

Danny looked around the room. "You know, me and your mum came to this café on our very first date?"

Will frowned and looked around as if the café had suddenly materialized around him.

"Nothing's changed by the looks of it. Even the tablecloths are the same."

Will peeled his elbows from the sticky laminate.

"Think he was here then too," said Danny, tilting his head towards the man asleep at the next table.

Will smiled.

"I wanted to take her somewhere nice, but neither of us had any money. Well, I had a few quid but I'd lost it all trying to win something from one of those grabber machines." Danny made a claw with his hand and lowered it onto Will's head. "You know that battered old teddy bear in your bedroom? The one that Mum gave you? I won that for her. It probably cost me ten times more than it would have cost to buy in a shop, but it got me a kiss at least."

Will looked like he'd just seen a pigeon get hit by a truck, not fatally but enough to leave it mangled by the roadside.

"Too much information?" said Danny.

Will nodded emphatically.

"Sorry about that." Danny took a sip of Coke. "You might not know this, but your mum knew all about you before you were even born."

Will frowned as he tried to make sense of what Danny had just said.

"It's true. She told me everything, right there at that table thirteen years ago." He pointed to a vacant table in the corner.

Will turned to look.

"We were chatting about stuff, just getting to know each other, you know? Favorite colors, favorite films, favorite ways to upset our teachers, that sort of thing. At some point she asked if I ever wanted kids, which I wasn't really expecting. I mean, it's a pretty serious question for a fifteen-year-old, right? She just wanted to see how I'd react, but I didn't know that at the time. She always liked to mess with people, didn't she?" Will nodded. "Remember when she told that Jehovah's Witness she was a devil worshipper?"

Will laughed. He drew a circle on his forehead with his finger.

"That's right, she even drew a pentagram on her forehead. Poor guy didn't know what to say. We never saw him again, though, did we? Anyway, so she asks if I want kids, and I say yes. I hadn't given it much thought, but I figured that was probably what she wanted to hear. Then I asked her the same question, and she nodded and said one word. You know what she said?"

Will shook his head. He leaned forward without touching the tablecloth.

"William. That's what she said. I didn't know what she meant so I asked her to explain, and she said, 'That's his name. William.' And then she went on to describe everything about you as if you already existed—as if you were sitting right there with us at the table. Blond hair, blue eyes, big feet, handsome as hell. She even knew you'd have a birthmark on your arm."

Will admired his birthmark as if it weren't a natural phenomenon but a gift his mother had given to him.

"Crazy, right? I laughed when she said it, but I stopped when I saw how serious she was. You know that look she gets. Anyway, you arrived a couple of years later, and you were everything your mum said you'd be, but even more handsome, of course. It was the most amazing thing. It was like you'd always been a part of her, right from the beginning, you know? And so even though she's not here anymore, she's not gone really, because now she's a part of you. She's in your smile, she's in your eyes, she's in the way you always used to pronounce the silent *l* in *salmon*. I see her every time you hold your toothbrush with your little finger sticking out like the queen when she's drinking her tea. I see her when you pout in your sleep like you're being told off in your dreams. I see her whenever you use your knife and fork like a rightie, even though you're left-handed, and I definitely saw her when you were tearing up that dance floor back there. I see her every time I look at you, mate, and that's why I could never, *ever* forget your mum. Because as long as you're here, she'll always be here too, you know?"

Will nodded slowly, his eyes so glassy that Danny could see his own reflection in them. He took some napkins from the dispenser as the waitress arrived to clear their plates.

"Food that bad, was it?" she joked.

"He's just got something in his eye," said Danny, winking at Will. "Want to grab some ice cream on the promenade?"

Will's face brightened at the mention of ice cream.

"Thought you might." Danny downed his Coke and returned his wallet to his pocket. "Come on, let's—*uuurrrp!*" He slapped his palm over his mouth, but the burp had already escaped. Will burst out laughing.

"Sorry about that," said Danny, his face turning redder than the plastic lobster trapped in the fishing net above him. "How embarrassing. Come on, let's—"

"Uuurrrp!"

Danny looked up to find Will grinning mischievously.

"That's not funny, Will."

Will's grin faltered at the sound of Danny's Dad Voice.

"Don't you know it's rude to burp louder than your father?" said Danny. He let rip with another burp, this one deliberate. Will giggled and did the same, and Danny responded with a belch so loud that it woke the old man asleep at the next table.

"Sorry," said Danny, raising a hand in apology. "It's the Coke. It's very gassy." He rubbed his chest for emphasis and tried to keep a straight face, but he burst into laughter the moment he saw Will sniggering at him from across the table. "Come on," he said. "Let's get out of here."

CHAPTER 22

He wasn't sure when it had happened exactly, but as Danny took a bow and thanked the crowd for their time, their applause, and, most importantly, their money, it occurred to him that he had, at some point since purchasing the panda suit, started to enjoy his newfound profession. He was commanding bigger crowds than ever before and he was making more money than anybody else in the park. He was still putting every ounce of energy into his performances, and he was still taking a sweat-soaked costume home every night, but it no longer felt like work because it no longer felt like a chore. He'd spent his entire adult life laboring on building sites without even the slightest recognition for a job well done. He'd spent so many years taking orders from people who couldn't order a pizza without screwing it up. He'd seen more than one colleague get seriously injured and each time wondered if he'd be next, especially during the winter when everybody's minds were as numb as their hands and accidents became routine. And now here he was, being applauded and cheered by scores of people for dancing in a park, without anybody telling him what to do or how to do it (apart from Krystal), where the only real threat to his well-being came from children trying to hug him and inadvertently head-butting him in the testicles.

Thanks to the panda, he'd also started to feel increasingly closer to Liz. Dancing had been such a big part of her life and now, unbelievably, it was a big part of his. Even though it was far too late for them to enjoy those lives together—to practice together, to watch *Dirty Dancing* together, to tear up the town together—doing something his wife loved so much made him feel as if he understood her just that little bit more. Danny felt, in the strangest way, that he knew his wife even better now than he did when she was alive. It was almost as if the panda had ceased being a costume and become like a medium, holding Danny's hand in one paw and holding his wife's in the other and connecting them in a way that he never could have imagined.

On top of all this, Will, while still not talking to him at home, was making a genuine effort to close that once cavernous but now almost leapable gap between them. Yesterday he'd woken up earlier than usual to help Danny make breakfast, a small miracle given how averse he was to mornings, and today he'd given his dad an unprompted hug before heading off to school. All in all, Danny had to admit to himself that for the first time in a long time, life felt good. In fact, aside from the ever-present fear of Reg, life felt better than good. Life felt great. Only when he saw Will clapping along with the rest of the crowd did he suddenly remember that his landlord wasn't the only tricky situation he had to resolve.

"I didn't know you could dance," said Will as the rest of the crowd trickled past him.

Danny took a seat on the bench and fished out his notepad and pen.

Pandas can do loads of amazing things.

"Oh yeah? Like what?"

Danny thought for a moment.

We can go invisible.

"No, you can't."

Yes, we can.

"I've never seen an invisible panda," said Will.

Exactly! wrote Danny.

Will rolled his eyes and sat on the bench beside him.

"Mo says pandas poop up to fifty times per day. That's pretty amazing."

It's true. We spend a fortune on toilet paper.

Will laughed.

Who's Mo anyway? wrote Danny.

"My best friend. His name's Mohammed, but everybody calls him Mo. He's, like, an animal expert. Did you know that a group of pandas is called an embarrassment? He told me that too, but I don't know if it's true."

It depends how much we've had to drink, wrote Danny.

Will smiled. "Who taught you to dance like that?"

A pole dancer called Krystal, wrote Danny, sure that Will wouldn't believe him anyway.

"Very funny."

I'm serious. She taught me after I rescued her bathrobe that was stolen by a wizard who can set things on fire with his mind.

"I might only be eleven, but I'm not stupid."

Eleven? I thought you were at least twenty-four, wrote Danny.

"I wish." Will laughed.

No, you don't. Keep being eleven for as long as possible.

"How old are you?"

Eighty-four in panda years.

"Well, you're a really good dancer for an eighty-four-year-old," said Will.

Danny put his paws together and gave a little bow of gratitude.

"My mum used to be a dancer. She was really good too."

What kind of dancer was she? wrote Danny.

"Every kind. She could dance to anything, even if there wasn't any music." He pulled out his phone. "Here," he said, showing the screen to Danny. "This is her."

Will pressed play and Danny watched as Liz danced alone in a spacious room with wooden floors and high ceilings reminiscent of a school assembly hall. He'd never seen the video, and the fact that he didn't even know where it was filmed caused the stitches of his soul to unravel slightly.

"She worked at a school," said Will, as if reading Danny's mind. "Sometimes she used to practice there, when nobody else was around."

Danny nodded, his bleary eyes fixed on the video as he took in every detail. What Liz was wearing. How she was moving. The way she brushed her hair from her face. The way she laughed and pretended to be mad when she realized that Will was recording her. The way she covered the camera with her palm before the video came to an end. Danny had plenty of videos of Liz, but seeing one for the very first time—one he didn't even know existed—momentarily made him feel as if she were still alive somehow, as if she hadn't died at all but slipped through a crack in space and time and ended up trapped in the video he was now watching. He wanted to replay it, and then replay it again, and keep replaying it until the battery eventually ran out; but suddenly aware that he was staring at the screen despite the video having already ended, and unsure of how long he'd been doing this exactly, he handed back the phone and scribbled something into his notepad.

Most people could live forever and not have that kind of talent, he wrote. Will smiled and nodded. A dog started barking somewhere nearby and they sat in silence for a moment while they watched a rowdy Jack Russell straining on its lead as it tried to pick a fight with an anxious pit bull.

Tell me something about your mum, wrote Danny.

Will shrugged. "Like what?"

Like anything.

Will stared across the park at something only he could see.

"She had these moles on her arm that sort of made a star if you joined them up with a pen. She used to let me do it for fun, but one time I did it with a permanent marker by accident and it took her ages to wash it off. And she was really good at crosswords, especially those cryptic ones with the clues that are super confusing. She was always trying to solve them, even when she wasn't looking at them. Sometimes we'd be having dinner or we'd be in the supermarket or something and she'd suddenly shout out a random word like the name of a country or the color of a certain type of horse. One time we were on the Tube and she shouted out, *'Leprechaun!'* and there was this really short woman sitting opposite us and she started yelling at Mum because she thought she was talking about her. It was pretty funny. And she always smelled of oranges because that's what her favorite hand cream smelled of. I have one of the empty jars and it still sort of smells like she did. I don't open it much because I don't want the smell to disappear, but the wardrobe in my bedroom has sliding doors and sometimes I sit in there and open the jar and the smell gets trapped. If I close my eyes, it's like she's right there with me."

Will flicked the loose button on his sleeve back and forth. Danny took the silence as his cue to say something, but Will continued before he could write anything.

"She drank, like, ten cups of tea a day, and she always put two tea bags into the cup because she liked it really strong, even though it tasted gross. She used to laugh at my dad because he couldn't drink it, even though he's a builder and builders are supposed to drink really strong tea apparently. She couldn't drink mint tea, though, or eat anything with mint in it, because it made her sneeze, and when she did she sounded like a mouse, or that's what my dad said, but my mum always said that mice don't sneeze. Mo says they do, though, but only when they're ill. She was left-handed too, like me. We have a left-handed pair of scissors at home, and me and Mum always used to laugh when Dad tried to use them because he couldn't. Oh, and her favorite color was yellow. She had lots of yellow stuff, like shoes and clothes and things. Even the scissors are yellow. She sometimes wore this yellow dress that made her look like sunshine, even when it was raining. I don't know where it is now though."

Danny nodded. He knew exactly where it was because he had picked it out for Liz to be buried in. He thought about that day and how surreal it had felt to be going through her wardrobe in search of a suitable outfit for such a horrendous occasion, and he recalled how it had rained so heavily and for so long after the funeral that it seemed as if what Will had said was true, that Liz was the sunshine and that they hadn't just buried her that day but also any light that was left in the world.

What's your favorite color? wrote Danny, realizing with no small amount of shame that he didn't know the answer.

"Guess," said Will, nudging his green schoolbag with his foot. "Mum said I could paint my room green too. And she said I could get a bunk bed, not one of those with two beds but one with a bed on top and a desk and stuff underneath it. I never did though."

Danny remembered having this conversation with Liz shortly before she'd died, she telling him that Will's bedroom needed redecorating, and he telling her that they couldn't afford it and there was no point anyway because they were only going to have to change it all again when Will became a teenager and decided that bunk beds were for kids. They'd argued about it, but knowing now how little time they'd had left together at that stage—he couldn't remember exactly how long, but it was only a matter of weeks—it made him feel almost painfully empty to think he'd ever wasted a second of that time quarreling.

You don't like your room? wrote Danny, already knowing the answer.

"It's got the worst wallpaper ever."

Don't tell me. Thomas the Tank Engine?

"How did you guess?" said Will sarcastically.

You should tell your dad. Maybe he can help.

"He already said no when my mum mentioned it."

Maybe he's changed his mind.

Will shrugged. He peeled a scab of paint off the bench.

You still not talking to him? wrote Danny.

Will shook his head.

Do you think you ever will?

"I don't know. He's being . . . weird at the moment."

Weird? wrote Danny, the word stinging after all he'd done recently.

"Yeah," said Will, "but, you know, in a good way. Like, he made me pancakes for breakfast the other day, and he never does that. I didn't even know he *could* do that. And he took me to Brighton at the weekend, and when we were there he played this dancing game with me and at lunch he talked a lot about Mum, and he never does *any* of those things. So yeah, it was weird, but a good kind of weird."

Maybe he's trying to be friends.

"Maybe."

Does he have any friends?

"Not many."

Then maybe he needs one, wrote Danny. *You don't make pancakes for somebody you don't like. Every panda knows that.*

"You *do* know you're not a real panda, right?" said Will, sliding off the bench and throwing his bag over his shoulder.

Have you ever actually seen a panda in real life?

"No."

Well then, wrote Danny.

Will smiled. "Whatever you say, panda-man-who-isn't-actually-a-panda. See you around."

See you around, not-talking-boy-who-actually-talks, scribbled Danny.

CHAPTER 23

Danny rang the doorbell and took a few steps back. He wasn't wearing the panda costume this time, but he nevertheless felt safer standing out of range of Ivana's broom or Ivan's chokehold. Yuri eventually answered the door. He was wearing a basketball T-shirt and the same combat pants that Danny had borrowed a few weeks earlier, and he was eating a family bag of Doritos that looked deceptively small in his massive hands.

"Oh, hi, Yuri," said Danny. "Wow, look at you. When are you going to stop growing?"

"Probably when I'm seventeen or eighteen or something. That's when most people stop growing, I think."

"Right," said Danny. He'd forgotten how literal Yuri could be.

"Want a Dorito?"

"What flavor?"

Yuri looked at the packet. "Blue," he said.

"Go on, then," said Danny, plucking a Dorito from the packet. "Is your dad home?"

"Yes," said Yuri. He didn't move.

"Can I see him?"

"Not from here," said Yuri, looking around as if to demonstrate how impossible it was to see his dad from their current location. "You should probably come in."

"Good idea," said Danny, squeezing past Yuri, who stood aside to let him through.

Finding the living room empty, he followed the noise that was coming from the kitchen. Ivan was standing with his back to the door, hunched over the work surface with an apron tied in a bow around his waist. Danny smiled, thinking his friend was playing some kind of joke on him. Ivan was not an "apron" kind of guy, but nor was he a "joking" kind of guy, which left Danny none the wiser as to what was going on. Only when he saw the open cookbook near Ivan's elbow did everything start to make slightly more sense. Ivan was baking, something that Danny had never seen him do before. He hadn't even realized his friend *could* bake, yet here Ivan was, deep in the middle of making what appeared to be a walnut cake—the same walnut cake that Danny had been eating for the last year or so. The fact that neither Ivan, Ivana, nor Yuri liked walnuts (something Danny had known since the day that Liz had unwittingly made her famous—now infamous—Waldorf salad for them) only confirmed his suspicions that Ivan was making the cake for *him*, and had been doing so since the very beginning. Danny felt an almost overwhelming desire to hug his friend, but not wanting to embarrass either of them, he slowly backed out of the kitchen and went to find Yuri, who was sitting in his bedroom playing on his Xbox.

"Hi, Yuri," said Danny, popping his head around the door.

"Hi, Danny," said Yuri, who appeared to be midway through a carjacking.

"Can you do me a favor and go tell your dad that I'm here?"

"Why?"

"I can't find him."

"He's in the kitchen," said Yuri, who still hadn't taken his eyes off the screen.

"I checked. I couldn't see him."

Yuri sighed and paused the game. He sat in silence for a moment.

"I can even *hear* him in the kitchen right now. Listen."

"I don't hear anything," said Danny, ignoring the obvious kitchen noises coming from down the hallway.

"He's definitely in the kitchen," said Yuri.

Danny sighed. "Can you just let him know that I'm here?"

Yuri shook his head in that way children do when adults are being weird. Then, rolling off his bed, he went to find his dad while Danny lingered in the hallway and listened to the conversation.

"Danny's here and he said he couldn't find you even though I told him you were in the kitchen so he asked me to find you and tell you that he's here."

Danny made a mental note never to ask Yuri to do anything ever again.

"He's in the kitchen," said Yuri as he shuffled back to his room, although his words sounded distinctly like "I told you so."

"Thanks," said Danny, realizing that the situation was now ten times more awkward than before.

"Danny!" said Ivan as he poked his head out of the kitchen. There was a panic in his voice that Danny had never heard before.

"Hi, Ivan. Is this a bad time?"

"No!" he said. "Is fine. Go sit in living room."

Danny did as he was told. Ivan appeared a minute later, easing his way out of the kitchen and closing the door behind him as if trying to contain an energetic puppy.

"So," he said, slightly out of breath as he slumped down opposite Danny. "How is life of dancing rat?"

"Complicated," said Danny, trying not to laugh when Ivan wiped his brow and left a streak of flour across his forehead.

"How is complicated? You dress like embarrassment, you dance, you stop the dance, you dress like normal person again. Simple."

"Did you know that a group of pandas is actually called an embarrassment? I just learned that today. An embarrassment of pandas."

"Not group," said Ivan. He pointed at Danny. "Just one. A group is worse than embarrassment. Is tragedy. Is like Chernobyl of pandas."

"Thanks for the support. And to answer your question, it's Will. He won't stop talking to me."

"So why you look sad? This is good, no?"

"No," said Danny. "It's not."

"Wait," said Ivan. "First you complain that Will, he never talk. Now you complain he talk too much. I am thinking you like to complain."

"Like I said, it's complicated. Look, remember when I told you about that time I saved him from those bullies? Well, now he keeps coming to the park to talk to me, but it's not *me* he's talking to. It's the panda."

"But you and panda are same person."

"Yes, but Will doesn't know that," said Danny.

"So tell him."

"It's not that simple. He's been talking *about* me. And about Liz. Stuff he'd never tell me if he knew that I was, well, me. And on the one hand it's incredible. I'm learning things about him that I never knew before. And he's *talking*, Ivan. He's finally talking! But if he knows it's me, then he'll never talk to me again." Danny sighed. "I don't know what to do. What would *you* do? Actually, don't answer that. What should *I* do?"

Ivan narrowed his eyes in the way he sometimes did when he was pretending not to understand English.

"I have an idea," he said after some clearly strenuous thinking. He leaned forward as if about to impart something highly confidential.

"If you make Will want to talk to you—to Danny, I mean, not to panda—then maybe he will stop wanting to talk to panda."

"Ivan, if I knew how to make Will talk to me, then I wouldn't be in this mess, would I?" he said.

"This is true," Ivan said.

"Actually," said Danny, chewing his lip, "maybe you're onto something."

"I am?" Ivan sounded surprised.

"Do you think Alf would let me have some of that scrap wood from the site?"

"I already ask," said Ivan, shaking his head. "I wanted to build the shelves for Ivana, but Alf, he say no." He pointed to three shelves cluttered with framed black-and-white family photographs that looked as if they'd been taken during a nineteenth-century blizzard.

Danny frowned. "*Those* shelves?" he said, pointing.

Ivan nodded.

"Those exact shelves?"

"Those exact shelves," repeated Ivan.

"I thought Alf said you couldn't take the wood?"

"He did. I take it anyway."

"How?"

"I go at night. Is easy. I show you."

"You sure? Ivana would never forgive me if you went back to prison." The words came out before Danny even realized what he'd said.

"Prison? What prison?"

"Nothing," said Danny, eager to move on, but Ivan stared at him in a way that suggested neither of them would be moving anywhere until he explained. "Wherever you got those," he said, pointing at Ivan's tattoos.

Ivan looked at his arms and frowned before a rare burst of laughter shot out of him so suddenly that Danny gripped the chair.

"You think these are prison tattoos?" he said.

"They're not?" said Danny, surprised by the disappointment in his voice.

"I am family man, Danny, not criminal man. These are from Yuri, not prison."

"You let your son tattoo you?"

"No, I let man at tattoo shop tattoo me."

"I don't understand."

"Look," said Ivan, rolling up his sleeves and placing his inked-up forearms on the table, "one day after work, I am very tired and so I fall asleep in chair. When I wake I see Yuri has pen and is making art all over me. On my arms, even my face. He was very young, five or six maybe, and he is having so much fun that I do not want to stop him, so I pretend to sleep until he is finished. When I open my eyes, I see what he has made. It is most beautiful thing I ever see. I love it so much I go to tattoo man that same day and ask him to make it forever."

"Absolutely no disrespect, because that is, quite possibly, the sweetest thing I've ever heard, but, well . . . couldn't you have just taken a photograph?"

"For what? To keep on telephone? To sit in picture frame? A telephone you can lose. A photograph you can lose. But these?" he said, tapping his arms. "You cannot lose arms."

"That's not technically true—"

"Of course is true! How do I lose arm? I cannot drop arm down back of couch. I cannot leave arm in supermarket trolley. I cannot forget arm in back of taxi. Is impossible."

"You're right," said Danny. Now didn't seem the right time to talk about the ins and outs of dismemberment.

"These," said Ivan, smiling as he gently traced one of Yuri's illegible scrawls with his finger, "are with me forever."

"I feel a bit bad now," said Danny. "I always assumed you'd just, you know, murdered someone."

"I said the tattoos came from Yuri," said Ivan. "I never said I have not murdered anybody." He winked at Danny in a way that left him none the wiser as to whether his friend was telling the truth or not.

"We should probably go," said Danny, sliding from his seat and shuffling towards the door.

"One minute. I just need to do something in kitchen. For Ivana."

"Anything I can help with?" said Danny, already knowing the answer but wanting to see his friend's reaction.

"No!" said Ivan, his voice uncharacteristically high. He cleared his throat and tried again. "I mean, no. Is fine." He opened the door just wide enough to fit through, which was pretty much all the way.

"Ivan?" said Danny. Ivan turned to face him. "Thanks."

"For what?"

"You know."

"I don't."

"Just . . . thanks," said Danny.

Ivan frowned and shook his head. "You're being embarrassment," he said before disappearing into the kitchen.

The building site was encircled by a tall fence of wire mesh. The only way in or out was through two large iron gates that blocked the entrance, next to which sat a small hut with two guards playing cards inside. Powerful floodlights shone down from all four corners of the perimeter, their beams converging in the middle where most of the prefab offices were located. The areas directly beneath the lights saw very little illumination, and nobody saw Danny and Ivan lurking in the shadows.

"Here," said Ivan, handing something to Danny in the dark.

"Is this really necessary?" said Danny, unrolling the fabric to reveal a black balaclava. "We're stealing wood, not storming the Iranian Embassy."

"Cameras," said Ivan, swirling his finger around his head.

Danny pulled on the balaclava and swiveled it so his eyes aligned with the holes.

"Okay," said Ivan. "Ready?"

Danny took a few deep breaths and stretched his hamstrings like Krystal had taught him.

"Okay," he said. "Let's do this."

He grabbed two handfuls of mesh and began to scale the fence, or tried to, but the wire dug into his fingers and he struggled to gain any traction with his feet.

Ivan grabbed the back of his shirt and yanked him down. "What are you doing?" he said.

"What does it look like? I'm climbing the fence."

Ivan sighed and shook his head. He grabbed the bottom of the fence and peeled it up until a hole appeared big enough for both of them to fit through.

"Or we could do it like that," said Danny.

They skulked across the site, sticking to the shadows wherever they could, until they reached the area officially known as the Shit Tip, where large yellow skips containing various discarded materials were lined up side by side. One was overflowing with broken bricks and rubble, another contained random pieces of plastic tubing and empty cement bags, and the last one had various offcuts of timber poking from it.

"Okay," said Ivan, making a cradle with his fingers and hoisting Danny up by his foot. "You find wood, I keep lookout."

Danny rummaged through the skip as quietly as possible, removing any suitable planks and handing them down to Ivan, who gently stacked them on the ground. Once a pile reached a certain size, he bound it with rope at each end and connected them to make a handle in the middle. Halfway through their fourth and final pile, Ivan suddenly froze and stared off into the distance.

"What's wrong?" whispered Danny.

Ivan shushed him, his eyes fixed on something in the darkness. "Somebody's coming!" he said.

"Shit! What do we do?" hissed Danny.

Ivan pulled a plank of wood from one of the bundles. He weighed it in his hand and slapped it into his palm like a bat.

"No, Ivan! No violence!"

"You have better idea?"

Danny frantically looked around. "Hide the wood and get in the skip!" he said.

"What?"

"I said get in here! Hurry!"

Seeing a torchlight rounding the corner, Ivan grabbed the bundles of wood and chucked them out of sight. Then, grabbing Danny's outstretched hand, he almost pulled him out of the skip as he scrambled into the metal container.

"What now?" he said, but Danny was already smothering himself with various planks and offcuts. Ivan did the same, burying himself as deep as he could and then covering his massive frame with an even larger piece of chipboard.

"I think it was coming from over here." One of the security guards appeared seconds later with his partner close behind him. Danny and Ivan closed their eyes as the man panned his torch across the skip.

"Probably just a rat," said the second security guard, who looked like a shorter and rounder version of the first.

"It must have been a fucking big rat to make that kind of ruckus."

"Rats can grow pretty big," said the shorter one. "I once saw a rat as big as a dog."

"Bollocks."

"Seriously. Not like a Great Dane or anything, but it was still massive."

"How big?" said the taller one.

"I don't know. Size of a bichon frise maybe. I saw it in Leicester Square, dragging a magpie into a bush. Poor thing wasn't even dead."

"A bichon frise isn't very big," said the taller one, clearly not convinced. "It's barely as big as a cat."

"Well, whatever. That's how big it was."

"So what you're saying is that you once saw a rat the size of a cat."

"No. It was the size of a dog," said the shorter one.

"Okay, it was the size of a dog that's the size of a cat."

"Yeah."

"So it was the size of a cat," said the taller one.

"No, it was the size of a dog."

"Jesus fucking Christ. Listen, if a rat is the same size as a dog, and the dog is the same size as a cat, then the rat, being the same size as the cat-sized dog, is therefore also the same size as a dog-sized cat. Correct?"

"I don't know, you're confusing me now."

"Okay, look. Imagine my left fist is a cat. And my right fist is a dog."

"Don't threaten me, Stu."

"I'm not threatening you!"

"Then get your fists out of my face!"

"Or else what?" he said.

"Or else I'll knock your bloody block off!"

Suddenly Ivan burst from the skip like a whale breaching the waves. The guards screamed and even Danny whimpered as planks and timber rained down around them while Ivan proceeded to thrash about like a loose tarp on a lorry.

"Run!" yelled the guard, but his partner had already gone, his torch throwing shadows all over the place as he sprinted off into the darkness.

Ivan continued his bizarre performance for another minute or two before finally grinding to a halt. Breathing heavily, he pulled off his balaclava and used it to wipe the sweat from his brow. Danny backed away, concerned that any sudden movements might trigger another episode.

"Sorry," said Ivan.

"What. The fuck. Just happened?" said Danny.

"Is not my fault. Is the wood bug."

"What . . . wood bug?"

"*This* wood bug!" said Ivan. He aimed a trembling finger at something that on closer inspection turned out to be a wood louse.

"A wood louse?" said Danny, ripping off his balaclava. "You almost made me shit my pants because of a fucking *wood louse*?"

"It bite me!" said Ivan, pointing to a perfectly unbitten forearm.

"Not unless you're made of wood, it didn't!"

"Well . . . it was thinking to bite me! I could see it in his eyes."

"I can't believe the great Ivan Shevchenko is scared of wood lice," said Danny as he gently rubbed his heart.

"Ivan is scared of nothing!" said Ivan.

Danny's eyes widened. He pointed at his friend. "Ivan, don't panic, but I think there's one on your shoulder!"

Ivan resumed his frantic dance, running in circles and swatting his body until he saw Danny laughing.

"What were you saying?" said Danny.

Ivan frowned. "I was saying, good luck carrying wood on your own," he said, tramping off across the building site.

"Wait!" said Danny, his smile wilting. "Ivan! Come back! It was a joke!"

CHAPTER 24

Danny woke up the following morning to find Will and Mo playing video games in the living room.

"Video games?" he said. "On a beautiful day like this?"

The boys didn't respond, their eyes fixed on the screen as they ran around trying to murder each other with chain saws.

"You kids should be outside, not glued to the TV."

"It's raining, Mr. Malooley," said Mo without looking up. Danny looked at the rain drumming against the window.

"That? That's nothing. And anyway, a bit of rain never hurt anybody."

"My Uncle Faisal drowned in a flood," said Mo.

"Oh. Right. Sorry to hear that, Mo. But look, the sun's going to come out any minute." Danny pointed at the window, ignoring the slate-colored clouds that stretched for as far as the eye could see.

"I don't know. It's raining kind of heavy. We might get sick or something."

"It's important for kids to get sick, Mo. It helps to strengthen their immune systems. And if you do catch a cold, you don't have to go to school, so you know, win-win, right?"

Mo and Will didn't look convinced. Danny resorted to plan B.

"Fine," he said, pulling out his wallet and taking out a tenner. "Here. Go and enjoy yourselves."

Mo looked at Will. Will shook his head. Mo looked at Danny.

"No offense, Mr. Malooley, but I think our lives are worth a little more than that."

"Lives? You're not going to die out there, Mo."

"That's what my Uncle Faisal thought," said Mo. Danny sighed.

"Okay, okay." Danny fished another ten-pound note from his wallet and offered them both to the boys.

Mo looked at Will. Will nodded.

"And don't come back until the evening!" shouted Danny as they ran off to get their raincoats.

Once they'd gone, Danny called Ivan, who arrived with a toolbox in one hand and a foil parcel in the other.

"From Ivana?" said Danny, smiling.

"What is this? KGB interrogation?" Ivan grumbled as he pushed his way past Danny.

They carried the bundles of wood upstairs from the disused garage they'd stored them in (but only after Danny had reassured Ivan that the planks were free from flesh-eating wood lice) and dumped them outside Will's door. Then, comparing notes on how best to proceed, the two men got down to business. They worked from midmorning until early evening, only stopping briefly for lunch and another time when Ivan had to explain to Ivana why he'd decided to spend his Saturday with Danny instead of taking her shopping at Westfield, a conversation that Danny was reluctantly dragged into when Ivana demanded to hear his voice as proof that Ivan was indeed where he said he was and not with the woman from flat fifty-four (who, according to Ivan, had simply locked herself out in her bathrobe and knocked on his door for assistance, which perfectly explained why Ivana found the woman seminaked in their house but didn't do much to explain why Ivan was also half-dressed).

The task proved harder than expected, not least because the room was so small and Ivan was so big and neither of them really knew what they were doing. They'd only just finished by the time Will walked through the door with a small box of leftover pizza in his hand. He found the two men in the kitchen, their faces and clothes speckled with paint and their cups of tea garnished with sawdust.

"All right, mate?" said Danny. Will stared at them from the doorway the way a train conductor stares at a student card he knows is fake but can't quite prove it. "You survived, then?"

Will nodded, but the frown remained.

"Is that pizza?" said Ivan, pointing at the box. Will nodded again and offered the box to Ivan, who slapped one slice between the remaining two and shoveled the pizza sandwich into his mouth.

Will pointed at Danny's clothes and turned his palms towards the ceiling.

Danny smiled. "All will be revealed," he said, putting down his tea. "But first you've got to close your eyes."

Will did as instructed, and Danny covered his eyes to stop him from peeking. He carefully ushered him out of the room and guided him through the flat until they came to a halt in the doorway of Will's bedroom.

"Okay," he said, taking his hands away as Ivan appeared behind them. "You can look now."

Will opened his eyes and stared at his room before closing them again and repeating the process. Danny watched him closely, nervously waiting for a sign of approval, but Will's face was inscrutable, like a poker player with a Botox addiction. The longer the moment went on, the more worried Danny became, first that he'd done the wrong thing and then that his son had connected the dots and sussed out his furry alter ego. It wasn't until he saw Will smile that his pulse began to stabilize.

Thomas the Tank Engine had suffered a fatal collision with a couple of strip knives with which Danny and Ivan had spent the morning hacking at the wallpaper. In his place were two lush coats of parakeet-green paint, the top layer still sticky to the touch (as Will accidentally discovered when he probed it with his finger). His old bed still occupied its place in the corner, but now it was closer to the corner of the ceiling than the floor, suspended almost six feet in the air by a sturdy structure of planks and posts that Danny and Ivan had spent the afternoon hammering together, much to the annoyance of the neighbors. A homemade ladder ran up the side of it, and beneath the bed was a study area complete with a desk, a table lamp, and a chair from the dinner table that Danny vowed to replace with a proper office chair just as soon as he could afford it. It wasn't the desk itself that caught Will's attention but the picture frame on the wall behind it. Staring back at him was a collage of pictures that Danny had spent the night compiling, all of them taken from Liz's photo albums, which hadn't been opened since the crash. Some of them were just of her, others of her and Will, and right there in the middle of the collage was a selfie Liz had taken of all three of them at London Zoo, smiling in front of the monkey enclosure with a photobombing spider monkey grinning through the bars behind them. Stuck above each of them were headings that Danny had made: *Mum* for Liz, *Dad* for himself, *Monkey* for Will, and *Will* for the monkey.

Will reached out and touched the picture, his fingers shaking slightly. He didn't see Ivan pat Danny on the back, and he didn't hear Danny step into the room until he felt his dad's hand on his shoulder. Will hugged him as hard as he could. Nobody spoke. Nobody needed to.

(HAPTER 25

Danny passed the next few days in a state of irritating cheerfulness. He whistled when he walked. He smiled when he talked. He sang in the shower. He said hello to strangers. He even tried to make peace with El Magnifico by flipping him some money following one of his performances, a gesture the magician responded to by fishing out the two-pound coin and lobbing it at Danny's head (which actually worked in Danny's favor, given that he'd only donated a pound to begin with). Only when he came home one day to find Reg and Mr. Dent in his living room did his good mood fizzle like a faulty firework.

"Where the fuck have you been?" said Reg from the couch, as if the three of them were flatmates and it was Danny's turn to cook dinner.

"All right, Reg," said Danny, trying to remain calm. He nodded at Mr. Dent, who was standing behind the empty armchair. "Dent."

"Was that from *Flashdance*?" said Reg.

"What?"

"The tune you were whistling just now. Good film, that. Jennifer Beals." He groaned like he'd just been kicked in the balls and enjoyed it. "You ever seen *Flashdance*, Dent?"

Mr. Dent frowned and shook his head.

"Dent isn't one for dancing. I used to be quite the mover, though, back in the day. Ballroom mainly."

"I didn't know that, Reg," said Danny, wincing at the thought of Reg doing the rumba.

"Yeah, well, we could fill a bathtub with the stuff you don't know. Sit down." He pointed to the empty armchair that Mr. Dent was standing behind.

Danny took a seat. "What can I do for you, Reg?"

"What can you do for me? What can *you* do for *me*? I'll tell you what you can do for me, Daniel. You can give me my fucking money."

"I will," said Danny. "Next week. Like we agreed."

"We didn't agree next week. We agreed today. Today is two months from when we agreed."

"Two months is next week," said Danny. He felt the need to loosen his tie despite not wearing one.

"Are you calling me a liar?" said Reg, his voice turning cold and sharp as an ice pick.

"Course not, Reg, I'm just saying—"

"Then you must be calling me a mug. Are you calling me a mug?"

"No, Reg—"

"Dent, is this mug calling me a mug?"

Mr. Dent shrugged.

"Please, Reg," said Danny. "Just give me until next week and I'll have your money, I promise."

Reg looked down at his lap. Only then did Danny notice the framed picture of Liz in his hands.

"You know, I never knew what she saw in you. She was a lovely girl, that wife of yours. She deserved better. She never should have shacked up with a useless fucker like you. I'm not trying to hurt your feelings here, Dan, I'm just telling it like it is."

"Thank you, Reg," said Danny, trying to sound like he meant it. "Very kind of you."

"See, when we're born we're all a bit like clay, ain't we, Dent?"

Mr. Dent nodded, even though he clearly had no idea what Reg was talking about.

"We come out as these ugly little gray lumps of nothing, and then life gives us color and molds us into different shapes and sizes until we eventually become who we are. But you know what your problem is?"

"No, Reg."

"You're still the same useless lump of shit that you were when you were born. But luckily for you, Dent here is something of a sculptor."

Before Danny knew what was happening, a rope appeared around his arms and waist, pinning him to the chair.

"What the—"

"But I have to warn you, Dan," said Reg as Dent tightly knotted the ropes. "He's a bit of a messy artist."

Another rope appeared, this one around his ankles. Danny began to struggle, but Dent had tied him firmly to the chair.

"What's going on?"

Dent loomed over Danny like a six-foot-five-inch stack of bad news. In his fist was the claw hammer.

"Reg, listen to me. I'll get your money, swear to God, but I can't fucking get it if you break my fucking legs!"

Reg slipped his arms into his crutches and waddled over for a better view.

"I'd like to tell you this isn't going to hurt," he said, "but from my own experience, I can honestly say that it's going to hurt like a bastard."

Danny yelled and thrashed in his chair like a pilot whose windscreen had just blown out. Mr. Dent raised the hammer and grinned like a kid playing Whac-a-Mole. He was just about to slam it down into Danny's trembling kneecap when Will suddenly burst into the living room.

"Leave him alone!" he yelled, planting himself between Danny and Dent and trying to make himself as big as his skinny frame would allow. Everybody looked surprised, but nobody more so than Danny.

"I thought he didn't talk," said Reg.

"He didn't," said Danny, smiling at Will despite his dire predicament.

"I liked you better when you were quiet," said Reg. Will didn't blink as he held Reg's bloodshot gaze. "Still, you got bigger bollocks than your old man, I'll give you that."

Reg sighed. Dent scratched his head with the claw and waited for further instructions.

"I guess it's your lucky day, Dan. There won't be another, so make the most of it. And the next time I see you, you better have my money. Otherwise," he said, turning to Will, "you won't be the only one that Dent's going to be denting. *Comprende?*"

"Got it," said Danny.

"Good lad," said Reg. He nodded at Dent. "Come on, Lurch."

The two men let themselves out while Will ran to the kitchen and returned with a pair of scissors.

"Thanks, mate," said Danny as Will cut him loose. He grabbed his son and pulled him close as soon as his arms were free. "It's good to hear your voice," he said, squeezing Will as tightly as he dared without breaking him.

"What's going on?" said Will.

"Nothing," said Danny, as if almost getting kneecapped were a semiregular occurrence.

"Tell me the truth, Dad, I'm not a baby."

"I know, mate, I know you're not a baby. And I'm sorry for treating you like one. I'm sorry for a lot of things, Will. I just . . . I haven't been myself since we lost your mum. Or maybe I *have* been myself, I don't even know anymore. All I know is that I should have been there for you and I wasn't, and I'm sorry for that. I'm so, so sorry. I know you've been upset with me and you have every right to be. *I'm* upset with me, but I'm going to make it up to you, I promise. There's nothing I can do to change what's happened, but everything's going to be different from now on if you'll give me a chance. I realize I haven't been a very good friend; I haven't even been a very good *dad*, but I want to be, and I think I can be if you'll just give me a chance. So what do you say? You think we can be friends?"

Will stared at the outstretched palm in front of him for so long that Danny's arm began to wilt. It was clammy by the time Will finally grabbed it.

"Friends," he said.

"Friends," said Danny. "And don't worry about this whole thing with Reg, it was just a big misunderstanding. Everything's fine. Everything's completely fine."

"Everything's fucked," said Danny, wiping the sweat from his brow as he stared at himself in the wall-length mirror. "If I don't win this competition then everything is absolutely, completely, well and truly fucked. Me. Will. My ability to walk unassisted. Everything. Fucked."

"True," said Krystal, handing him a towel and sitting on the step beside him. "I'll be fine though." She nudged him in the ribs, but Danny didn't smile.

"You know my son had to rescue me from having my knees turned into pâté yesterday? In my very own living room? I mean, what kind of thing is that for an eleven-year-old kid to see?"

"An important thing. Your son just learned a valuable life lesson."

"And what lesson would that be exactly?" said Danny with an empty laugh. "Don't grow up to be a massive failure like your dad?"

"I'm sure he already knew *that*," said Krystal. "But now he also knows that all landlords are total wankers. That's what they should be teaching kids in school, not that maths and science bollocks but practical, useful stuff, like, you know, how to get served at a crowded bar, and how to talk your way out of a speeding ticket, and how to rewire a plug, and how to identify a dodgy landlord. I wish somebody had taught me this stuff before I moved into my last place."

"Had a lot of plugs that needed rewiring, did you?"

"No, but I had a creepy fucking landlord whose brain needed rewiring. He used to let himself into my flat and steal my underwear when I was at work. Always the expensive stuff too. He kept it all in a drawer by his bed."

"How did you catch him?"

"One day he was working in the front garden, and when he crouched down to weed the flower bed, his shirt rode up and I saw he was wearing my crotchless knickers."

Danny almost choked on the bottle of water he was drinking from. He wiped his mouth and looked at Krystal.

"What?" she said. "They were a gift."

"Well, I wish that's all Reg wanted," said Danny. "He's more than welcome to steal my undies."

"You know, most people would probably think about moving if their landlord tried to cripple them," said Krystal. "Just saying."

"It's not that simple," said Danny.

"Course it ain't simple. Moving's a total pain in the arse, but I'm pretty sure it ain't nearly as painful as being kneecapped."

"I don't mean like that. I mean . . . I don't know, it's difficult to explain."

"Don't mean it's difficult to understand," said Krystal, waiting for Danny to elaborate. He sighed and played with the bottle cap while he tried to find the right words.

"It's just . . . me and Liz, we moved into that flat together. And, well, to me it's still *our* home. I know she's gone, but she's also still there in that flat somehow. I know it, I can feel her. I found one of her hairs the other day. It was right there on the couch, like she'd literally just been sitting there the second before I walked into the living room. Crazy, right? She's been gone for over a year, and then a piece of her appears out of nowhere. That's why I can't go. I know it probably sounds stupid, but I can't just leave her like that."

"It's not stupid," said Krystal. "I get it. But you've got to realize that she doesn't live there anymore, Dan. She lives in here now," she said, tapping Danny's temple, "and here," she said, patting his chest before wiping her hand on her shirt. "And she'll be with you wherever you decide to go, especially if it's somewhere that isn't owned by a fucking psychopath."

They sat in silence for a moment, their conversation hanging in the air like the disco ball above them.

"Well, it's too late now anyway," said Danny. "I can't move out, even if I wanted to. Not until I pay Reg what I owe, and I can't pay Reg unless I win this competition."

"And you can't win this competition unless you keep practicing, so on your feet, soldier, let's get to it. There's no way I'm letting Kevin beat you without a fight."

"Kevin?"

"El Magnifico," said Krystal, rolling her eyes. "Otherwise known as Ballsack McFuckface."

"Seriously, how did somebody like you and somebody like him end up together?" said Danny, partly out of curiosity and partly to postpone having to dance for another few minutes.

Krystal shrugged. "I guess I was just going through one of those phases."

"And what phase would that be?"

"The 'I need to date an arsehole magician who refers to his willy as his wand and insists you scream abracadabra whenever you orgasm' phase."

Danny cringed. "Did you really have to say abracadabra?"

"No idea," said Krystal. "He always finished before me."

Danny shuddered.

"What? You asked." She smiled and shook her head. "What can I say, I was young and stupid. He was looking for an assistant and it sounded like easy money. I wasn't planning on falling for him or anything, I just needed the cash. I didn't even find him attractive, but, well, things have a funny way of working out sometimes, don't they? And by 'funny' I mean not fucking funny in the slightest. You know that whole sawing-a-person-in-half trick? Well, I was the person he was sawing in half, quite literally as it turned out."

"You seem to have recovered pretty well," said Danny, looking Krystal up and down.

She laughed. "Okay, fine, not 'literally', but he definitely cut me in half emotionally. I put myself into that stupid box of his night after night, and you know what he was doing? He was going home and putting himself into Carla's box night after fucking night."

"Who's Carla?"

"My sister."

"Oh."

"Oh is right," said Krystal, smoldering so ferociously that Danny swore he could smell smoke. He tied and retied his shoelace while he waited for her to burn herself out.

"I didn't know you have a sister," he said.

"Had," said Krystal. "I *had* a sister."

"What happened to her?"

"She's dead," said Krystal.

Danny nodded solemnly. "I'm sorry."

"Dead to me, I mean," said Krystal. "Stupid cow works in a warehouse in Bracknell."

"I think I'd rather be dead, to be honest," said Danny.

Krystal smiled and looked around the room. "We did a show here once actually, me and Kev, back in the early days."

"Fanny's doesn't seem like the sort of venue for a magic show."

"It was an 'erotic' magic show," said Krystal. "That's how Kevin sold it anyway. Got to hand it to him, he always knew how to market himself. There was nothing erotic about it, of course, it was the same old bollocks we always did, the only difference being that this time I had to strut around in my underwear, which wasn't ideal, obviously, but it also wasn't half as bad as having to squeeze into that sparkly latex shit that Kevin usually made me wear. Fuck me, that thing was hot. It was actually quite nice to be up onstage without sweating like Satan's scrotum for once, so when Fanny offered me a job after the show—she said I could make five times more money working for her than I could working for, and I quote, 'that pointy-hatted twat'—I was sort of tempted, but I turned her down because I happened to be in love with that pointy-hatted twat at the time. But then I found out about him and the bitch of whom we do not speak and I thought, fuck it, I'll do a bit of dancing until something else comes along, and, well, here I am, five years later."

"Why don't you leave?"

"Because there ain't too many jobs out there that pay as well as this. And, to be totally honest, I enjoy it. I know it ain't the most glamorous job in the world, but Fanny is good to me, Suvi looks after me, and taking money from stupid fuckers is even more fun than spending it."

"Maybe I'm in the wrong job," said Danny.

"Well, you've definitely got the moves, but I don't know how you'd look in a thong."

"Then cherish that fact," said Danny.

"Too late," said Krystal, her lip curling. "I've already got mental images."

"How do I look?"

"Like my old landlord."

"Ouch," said Danny.

Krystal laughed. "Come on," she said, standing. "Let's get back to it. I need to dance this thought out of my head."

CHAPTER 26

Mr. Coleman picked a paper airplane off the classroom floor and screwed it into a ball.

"How is it," he said, throwing the defunct aircraft into the bin, "that man has managed to climb to the peak of Mount Everest, he's located the source of the Nile, he's trekked to both the North and South Poles, and he's circumnavigated the globe in a hot-air balloon, yet you lot still haven't figured out how to find your chairs without a map and compass?"

"How is this going to help me find my seat?" said one pupil, brandishing a maths compass. Everybody laughed.

"Just aim it at your desk and follow the pointy end," said Mr. Coleman.

"Look, Mr. C," said another student. He showed Mr. Coleman his iPhone. The boy had typed *my chair* into Google Maps. "No results," he said.

Mr. Coleman took the phone and typed *Wilson's brain* into the search bar.

"What a surprise," he said, returning the phone to Wilson. "No results there either. Come on, everybody, in your seats, now. The last person to sit down has to eat their lunch with me."

The classroom erupted in a flurry of movement as children scrambled to find their seats.

"I'm glad to see you're so eager to learn all of a sudden!" said Mr. Coleman.

He sat behind his desk and opened his glasses case.

"Atkins?" he said, hunching over the register.

"Present," said Atkins.

"Cartwright?"

"Here. I mean present."

"Jindal?"

"Present."

"Kabiga?"

"Present."

"Malooley?" Mr. Coleman glanced at Will and placed a tick beside his name.

"Present," said Will.

"Moorhouse?"

Moorhouse didn't answer, and Mr. Coleman didn't ask a second time. He stared at the register, his brow creased as if trying to remember whether he'd fed the cat that morning. Slowly removing his glasses, he looked up at the class. Every head in the room was turned towards Will, and even Mr. Coleman couldn't stop himself from staring.

"Will?" he said.

Will smiled. "I'm still present," he said as the rest of the class, especially Mo, continued to stare at him in disbelief.

Mr. Coleman nodded, too stunned to say anything. Then clumsily putting his glasses back on, he cleared his throat and returned to the register.

"Er . . . Saltwell?"

*　　*　　*

"Okay, people!" yelled Mo, taking over crowd control and ushering everybody back so Will could emerge from the corner he'd been backed into. "Give the boy some space!"

"Say something!" shouted one of the kids from the jostling crowd. Word about Will had spread around school, and now everybody had gathered to see if the rumors were true.

"Something," said Will. They all laughed.

"Say something else!" shouted another.

"Something else," said Will to more fits of giggles.

"Say 'She sells seashells on the seashore!'" shouted a girl from somewhere in the crowd.

"She sells seashells on the seashore," said Will.

"Say 'Supercalifragilisticexpialidocious!'" shouted someone else.

"Supercalifragilisticexpialidocious," said Will.

"Say '*Hěn gāoxìng jiàn dào nǐ,*'" said Gan, a Chinese boy from the year below.

Will laughed. "I can't say that," he said.

"Say 'My name's Will and I love attention because I'm a massive fucking loser with no friends,'" said Mark as he barged in with Gavin and Tony in tow.

"Leave him alone, Mark," said Mo as a nervous ripple spread through the crowd.

"Shut it, Mo. Little Willy here can speak for himself like a big boy now, can't you, Willy?"

Will said nothing, the urge to speak suddenly gone.

"Say it," said Mark. Will shook his head. Mark grabbed him by the collar and put his face so close to Will's that flecks of spit landed on his cheek.

"Say it!" he said.

Will sighed. "My name's Will and I love attention," he mumbled.

"Because?"

"Because I'm a massive loser with no friends."

"Did you hear that, fellas?" said Mark, turning to Gavin and Tony.

"Nope," said Gavin.

"Not loud enough," said Tony.

Mark grinned. "Say it again so everybody can hear," he said.

"My name's Will and I love attention because I'm a massive loser with no friends!" said Will, louder this time.

"Louder!"

"My name's Will and I love attention because I'm a massive loser with no friends!" yelled Will.

"And don't forget it," said Mark. He leaned in close and lowered his voice, but the anger remained. "You think you're so special just because your mum died? Boo-fucking-hoo. My dad died two years ago, but you don't see me acting like a baby, do you? You don't see me trying to get attention because of it like some fucking loser. So why don't you stop embarrassing yourself and grow up instead of being such a pussy." Mark let go of Will's collar and shoved him against the wall. "Come on, lads."

Gavin and Tony followed him through the crowd that had parted for them. Realizing the fun was over, everybody else began to shuffle away until only Will and Mo remained.

"I didn't even know he *had* a dad," said Mo. "I thought he was grown in a lab or something."

Will wiped the spit from his cheek and stared at Mark across the schoolyard.

Danny arrived at the gates just as the school bell rang. The main doors opened and the kids came racing out as if fleeing the scene of a crime, which some of them probably were.

With the competition less than a week away, Danny had skipped his session in the park in exchange for more time to practice at

Fanny's. He now knew every part of his routine by heart, and even though he still hadn't mastered all of it, or half of it, or even a small percentage of it, he was at least able to dance his way through it from start to finish.

Danny rubbed his groin and groaned before reminding himself he was outside a school. He'd decided to wear the panda costume during that day's rehearsals in order to get a feel for how the fabric responded to his movements (it chafed like hell, especially his inner thighs) and how easily he could breathe with the mask on (about as easily as he could breathe with a carrier bag over his head), and he could feel the sweaty outfit soaking through his backpack as Will crossed the road towards him.

"What are you doing here?" said Will.

"Sorry, mate, am I cramping your style?" Danny looked around, conscious that an unknown love interest might be somewhere nearby.

"I have no style to cramp," joked Will. "I just thought you'd be at work."

"I, er, got off early," he said, having momentarily forgotten about his supposed job at the building site in his eagerness to see his son. He patted his backpack for emphasis. "Got my work stuff right here. I thought that maybe we could, you know, do something."

"Like what?" said Will.

"Whatever you want."

Will thought for a moment. "Burger King?" he said.

"Burger King it is."

"Can I have a triple Whopper with cheese?"

"Can you *eat* a triple Whopper with cheese?" said Danny.

Will shrugged. "I don't know."

"Then I guess there's only one way to find out. Come on."

"Wait," said Will, pointing in the opposite direction. "Let's go this way."

"But Burger King's the other way," said Danny.

"Just follow me."

Will didn't tell him where they were going and Danny didn't ask, content to follow along and enjoy the simple yet uncommon pleasure of sharing a conversation with his son. Will told Danny about school while carefully omitting the fact that Mark was continuing to make his life miserable, and Danny told Will about work while carefully omitting the fact that he'd been fired almost two months ago. He was so distracted by their conversation that he didn't realize they'd walked into the park until Will paused at the spot where Danny usually performed as the panda.

"What's up?" said Danny.

"I wanted you to meet somebody," said Will, scanning the park.

"Who?" said Danny cautiously, already knowing the answer but still obliged to ask.

"This guy who dresses up like a panda. He dances too. He's really good."

"What have I told you about talking to weirdos in parks?"

"He's not a weirdo, he's my friend."

"All right, Danny," said somebody behind them. "Didn't recognize you without your fur on."

Danny could almost feel his brain cells scurrying around in panic as he turned to face Tim. Milton was perched on his shoulder sporting a blue turtleneck sweater.

"Day off today, is it?" said Tim.

"I'm sorry, but do I know you?" said Danny. He winked and hoped that Tim would get the hint. He didn't.

"Do you know me?" said the busker with a laugh that started out genuine and ended up nervous.

"I mean, have we met before?" said Danny, winking again.

Tim winked back despite having no idea what they were winking about. "I don't know. Have we?"

"No, I don't think we have," said Danny.

"Then I guess we haven't," said Tim. He winked again.

"How did you know his name, then?" said Will.

"What?" said Tim, noticing Will as if for the first time. "I didn't. I don't."

"You called him Danny."

"No, I didn't."

"Yes, you did."

"Oh," said Tim. "Yes. You're right. I did."

Will waited. Danny squirmed. Tim fidgeted. Milton licked his arse.

"I call everybody Danny," said Tim. "It's just, you know, how I roll."

"You call everybody Danny?" said Will.

"Yes," he said. "That's correct." He turned to Danny, his eyes pleading for help, but Danny seemed to have frozen, like his system had crashed and hadn't yet rebooted. Tim continued with his narrative, trying to dig his way out by digging even deeper. "It's a funny story actually," he said. "But also kind of sad. You see, Danny is what I used to call my father. He was called . . . Bernard. Like Bernard Matthews. You know, the turkey bloke. He wasn't Bernard Matthews, though, just to clarify. Anyway, when I was young I called him Daddy, like children do, but I couldn't pronounce the word properly because . . . I was born with a speech impediment. Which is gone now. Obviously. But back then, whenever I tried to say 'Daddy', it came out sounding like 'Danny.'"

Danny recovered himself enough to make a series of chopping motions to his throat, repeatedly mouthing the word *stop*, but Tim was far too engrossed in his fabricated life story to notice.

"One day, without warning, my dad walked out and left us." He snapped his fingers. "Just like that. Gone. It was heartbreaking."

Danny looked on in disbelief as Tim choked up for a second. "Nobody knew where he went. Unconfirmed sightings were reported from across the globe, from the Ural Mountains to the jungles of West Papua, but no proof was ever found. I never stopped looking for him, but as the years passed I started to worry that we might not recognize each other anymore, so whenever I saw somebody who I thought might be him, I'd go up to them and say 'Danny,' because that way, if it was him, then he'd know for sure that it was me, his son. And that, young man, is why I call people Danny."

Tim looked at Danny, clearly impressed with himself.

Will frowned, even more confused than before. "So . . . you're saying that you thought *my* dad might be *your* dad?"

Tim's smile faltered at the realization that his story wasn't quite as watertight as he thought.

"Yes," he said. "I mean, no. Maybe." He looked at Danny. "Are you?"

"No," said Danny wearily. "I'm not your dad."

"Whatever. I'm going over there," said Will, pointing to a nearby crowd. "What were you saying about talking to weirdos in parks?" he whispered on his way past Danny.

"What the hell just happened!" said Tim once Will was out of earshot. He wiped his brow with Milton's tail.

"You just told my son that you thought I might be your dad, and you're asking *me* what the hell just happened!"

"I'm sorry, I panicked. What was I supposed to do?"

"No," said Danny, rubbing the tension out of his neck. "I'm sorry. It's my fault. I should have told you earlier. My son doesn't know about the whole panda thing."

"Why not?"

"Because I'm a dancing panda bear, Tim. It's not exactly something to be proud of, is it?"

Tim gave Danny's shoulder a squeeze. "*I'm* proud of you," he said. "And so is Milton. He might not look it, but he is."

"Thanks, I think," said Danny. "Hey, sorry about your dad, by the way."

"What about him?"

"He didn't walk out on you?"

"If he did, then I've got no idea whose basement I'm living in. By the way," he said, pointing over Danny's shoulder, "you might want to keep your son away from El Magnifico."

"Shit!" said Danny, noticing Will in the crowd. "Thanks again, I owe you one!"

"This guy's really good," said Will as Danny ran over.

"Yeah," said Danny, trying to hide behind the people in front of him. "Come on, we should probably make a move."

"For my next trick," El Magnifico addressed the crowd, "I'm going to need two volunteers."

Danny hunched his shoulders and stared at his feet, totally unaware that Will's arm was twitching in the air beside him.

"It looks like we've found ourselves some victims!" said the magician, pointing to Will. Everybody laughed. "Make way, ladies and gentlemen!"

"Come on!" said Will, grabbing Danny's hand.

"Will, stop!" whispered Danny, trying to resist without making a scene.

"Well, well, well," said El Magnifico when he saw Danny reluctantly emerge from the crowd. "Who do we have here, then?"

"I'm Will, and this is my dad."

"Let's hear it for Will and his dad, everybody!"

A gentle patter of applause rose from the crowd.

"Now, before we begin, Will, do you have a mobile phone?"

"Yes," said Will, removing his phone from his pocket.

"Excellent. Would you mind holding it up so everyone can see?"

Will held out his phone and swept it across the crowd as if taking a panorama.

"Very good," said El Magnifico. He turned to Danny and grinned like he'd just stuck a KICK ME sign on his back. "And you, sir. What's your name?"

"Danny."

"Do you have a wallet, Danny?"

"I do," he said, removing his wallet from his pocket.

"And could you also show it to the crowd?" said El Magnifico. Danny did as requested. "Pay attention to the finer details, everybody. The clearly fake leather. The cheap stitching. The distinct lack of money inside. Thank you very much, Danny, you can put the wallet away now. Do you mind telling us what you do for a living?"

"He's a builder," said Will.

"A builder?" said the magician with pantomime exaggeration. "Really?"

"Really," said Danny, trying to murder El Magnifico with his eyes.

"Isn't that interesting," said El Magnifico, pretending to twizzle his drawn-on mustache. "And do you do anything else? A second job maybe? Bartender? Postman? Dancing panda, perhaps?"

"Nope," said Danny through his teeth. "I'm just a builder."

"If you say so." El Magnifico turned to Will. "Tell me, Will, do you trust your dad?"

Will nodded with more confidence than Danny would have expected.

"And you, Danny, do you trust your son?"

"Of course."

"Isn't that precious, ladies and gentlemen?" said El Magnifico. The crowd murmured in agreement. "But, Will, what would you say if I told you that your father was actually . . . a thief!"

"Seriously?" whispered Danny, thinking the magician was still upset about his missing robe.

"I'd say you were a liar," said Will.

"Well, your loyalty is admirable, Will, but if I'm a liar, why does your dad have *your* phone in his pocket?"

"He doesn't."

"Are you absolutely sure about that?"

"Yes," said Will. He patted his trousers, but the pockets were suddenly empty. "Wait, no. Where's my phone?"

"Danny, could you check your pockets please?"

Danny halfheartedly patted his trousers, sure that he'd find nothing out of the ordinary. Instead he found something bulky that hadn't been there a minute ago. He slipped his hand into his pocket and pulled out Will's phone.

"Will, is that your phone?" said El Magnifico.

"How did you do that?" said Will, taking his phone and staring at it as if his whole life had been a lie. A few people clapped. Others checked their pockets to make sure their phones were still there.

"Don't ask me. Ask your dad. What do you have to say for yourself, Danny?"

"You got me," said Danny, holding his hands up in mock surrender.

"Still trust your dad, Will?"

"Yeah," said Will. "Like, eighty percent." The crowd laughed.

"Well, he might not trust *you* when he realizes you've stolen his wallet."

"I haven't," said Will, turning out his pockets. "See?"

"Danny, do you have your wallet?"

"No," said Danny, frowning as he checked and rechecked his trousers.

"Will, could you show us what's in your schoolbag?"

Will slipped his bag from his shoulder and began to root around in it. "It's just books," he said. "And a pencil case. And an old sock I didn't know was in there. And an even older apple." He held up the withered fruit to the sound of laughter. "But nothing— Oh, wait."

Will's hand slowly emerged holding Danny's wallet. The crowd applauded. Will looked stunned. Even Danny was impressed.

"Will, can you do me a favor and go through the wallet to confirm that it is in fact your dad's? Look for an ID card, something like that."

"Yep, it's mine all right," said Danny, laughing nervously as he finally understood what El Magnifico was up to. "No need to verify it."

"I found a bank card," said Will as he rummaged through the wallet.

"Anything else?" said El Magnifico.

"You've made your point!" hissed Danny, but El Magnifico just grinned.

"And a Nectar card," said Will.

"Keep looking," said El Magnifico.

"Oh yeah, here's something," said Will. He held up Danny's street performer's permit.

"What is it?" said the magician, literally rubbing his hands with glee. "Read it to me."

"It's—"

Before Will could finish his sentence, Milton leapt onto the table and attached himself to El Magnifico's face. The man screamed and dropped to the floor while the crowd, believing the assault to be part of the performance, began to film the event on their phones. Taking advantage of the chaos, Danny grabbed both the wallet and the permit and stuffed them into his pocket. Then, noticing Tim on the fringe of the crowd, he flashed a grateful thumbs-up before ushering Will away from the scene.

* * *

It was late by the time they got home that night. As promised, Danny had taken Will to Burger King where, also as promised, Will had been treated to a triple Whopper with cheese, a purchase that effectively answered the question of whether or not he could eat one by himself (he could, much to Danny's amazement, and also his disappointment, having not bought anything for himself assuming that the leftovers would be more plentiful than the solitary slice of gherkin he ended up with). Afterwards, on their way past the cinema, Will had dropped several not-so-subtle hints about how much he wanted to see the latest installment of the seemingly never-ending *The Fast and the Furious* franchise and how it was going to be gone from the cinema soon and how it was so much better to watch a film like that on the big screen and how he couldn't watch it on his own because he didn't have the money and it was a 12A anyway so he couldn't see it unless he was accompanied by an adult or unless he looked old enough to sneak in (which they both knew he didn't) until Danny had eventually capitulated.

"Don't tell anybody that I let you stay out so late," said Danny as he tucked Will into bed.

"It's not *that* late," said Will, unsuccessfully stifling a yawn.

"Then why are you yawning?"

"I'm not," said Will, his eyes closing despite his best efforts to keep them open.

"Well, I am," said Danny, yawning into his palm. "Get to sleep."

Danny turned the light out and began to close the door behind him.

"Dad?"

"Yes, mate?" said Danny, pausing with his hand on the handle.

"I had a good day today," said Will.

Danny smiled. "Me too," he said, but Will was already too far gone to hear him.

In the living room, Danny opened his backpack and almost gagged at the stench. Holding his breath, he yanked the costume out so quickly that everything else came tumbling out with it. He threw the costume into the wash and wearily waited for the cycle to finish before crawling into bed, completely unaware that his notepad was lying on the living room carpet.

CHAPTER 27

Will woke to the sound of banging. He thought it was the front door at first and momentarily panicked, worried that Reg and Dent had returned to finish what they'd started the other day, but the noise seemed to be coming from the kitchen.

Climbing down from his bunk, he pulled on his school trousers and opened the bedroom door a crack. He could hear singing as well as banging now, and it wasn't until he crept down the corridor and peered around the fridge that he figured out what all the commotion was about.

Danny was dancing around the kitchen with his earphones in and his back to Will. He plucked the toast from the toaster and tossed each piece in the air before catching them one by one on a plate. Only when he pirouetted his way to the fridge to grab the margarine did he notice Will watching him from the doorway.

"Hi, mate," he said, quickly pulling his earphones out. "You're up early."

"Were you just listening to the *Dirty Dancing* soundtrack?" said Will, his lips twitching with barely suppressed laughter.

"What?" said Danny, fumbling with his iPod as he tried to turn the music off. "No. I mean, I don't know. Maybe. Wait, how do you know *Dirty Dancing*?"

"I've seen it with Mum like a hundred times. I didn't think it was your kind of thing."

"It's not. I didn't even know that song was from *Dirty Dancing*. I've never even heard it before."

"Then how do you know all the lyrics?"

Danny opened his mouth to speak and then closed it again. He held up a piece of toast.

"Jam or Marmite?" he said.

Will smiled and rolled his eyes. "Jam," he said, shuffling into the living room and taking a seat at the table. "I hate Marmite."

"Since when?"

"Since forever," said Will, absently playing with a two-pence piece that he found on the table.

Danny made a mental note about the Marmite. "What classes you got today?" he shouted, eager to change the subject.

"History, science, English, and maths," said Will, spinning the coin between his fingers.

"Four of my worst subjects."

"Every subject was your worst subject," said Will as the coin spun from the table and landed on the floor by his chair.

"That's not true. I got a C-minus in art."

"That's what I mean." Will leaned down to retrieve the coin but found Danny's notepad instead.

"My art teacher was this woman called Miss Black. She was terrifying. Did I ever tell you about her?"

Will didn't respond, too busy flicking through the pages and trying to figure out what all the columns and calculations meant.

"She had this glass eye that she sometimes took out and washed in front of the class. I still have nightmares about her."

Will stopped flicking. He stared at the words in front of him. Words he'd seen before. Words he'd spoken before. *He doesn't know anything about me.... Tell me something about your mum.... Pandas are great listeners.* He scanned the pages, trying to make sense of what he was looking at, but the more he read the more he felt like the butt of a joke he didn't understand.

Danny emerged from the kitchen with a cup of tea in one hand and a plate of toast in the other.

"One time she sneezed so hard that her eye—" Danny fell silent when he saw what Will was looking at.

"This is how you knew," said Will, his eyes fixed on the words in front of him.

"Will, I—"

"Going to Brighton. Making pancakes. Decorating my room. You made me tell you everything."

"It's not like that," said Danny. He put Will's breakfast down and took the seat opposite. "I wanted to tell you, mate, really, but—"

"This is why the man at the park knew your name, isn't it?" said Will. Danny started to answer, but Will cut him off. "You tricked me," he said. "You lied to me."

"I didn't lie about anything, Will. And I didn't trick you either. *You* started talking to *me*, remember? What was I supposed to do? Ignore you?"

"You could have told me. But you didn't. You just let me keep talking like an idiot."

"Will, you hadn't spoken for over a year. I didn't know if you were ever going to speak again, so when you started I—"

"I didn't talk because I didn't want to talk!" yelled Will, his cup of tea rippling as he slammed the table with his palms.

"I know, mate," said Danny, showing his own palms in surrender. "I know. You're right. And I'm sorry. I'm really, really sorry."

"Why were you even wearing a panda suit? Why were you danc- ing in the park? Why aren't you going to work?"

"It *is* my work," said Danny. He sighed. "Alf fired me a couple of months ago and, well, I've been doing the panda thing ever since."

"And you didn't think that was something you should tell me?"

"I didn't want you to worry."

"Worry?" said Will. He laughed, but there was no humor in it. "I came home the other day to find you tied to the chair and Mr. Dent about to whack you with a hammer. Don't you think I found *that* slightly worrying?"

"I'm sorry you had to see that, mate, really, but everything's going to be fine, I promise."

"How can you make a promise when you can't even tell the truth?" Will threw the notepad at Danny. "I thought I could trust you! I thought you were my friend!"

"You *can* trust me! We *are* friends!"

"No, we're not!" Will got up and grabbed his schoolbag. "Mum was my friend, not you!"

"Will, wait, please," said Danny, following his son to the front door.

Will paused, but he didn't turn around. "You know what I wish?" he said. He wasn't shouting anymore, but something about his tone made the yelling suddenly preferable.

"What, Will? What do you wish?" said Danny, but he already knew the answer, because not a single day went by that he didn't think the same thing. "You wish it was me, don't you? That I had died instead of her."

"No," said Will, turning around to face Danny. "I wish it was *me.*" He jabbed himself in the chest. "I wish that *I* had died, *with* Mum, because I'd rather be dead with her than stuck here with you." He yanked the front door open and slammed it shut behind him.

Danny didn't know if Will had chosen those words deliberately, but they cut him more than when Liz's father had spat them at the funeral.

Now he's stuck without a mother.

Now he's stuck with you.

"Hold on, what?" said Mo as he fiddled with his hearing aid. "I think this thing is broken. What did you just say?"

"I said my dad's a dancing panda," said Will. He kicked a stray tennis ball across the crowded schoolyard.

"Maybe it's the batteries or something. It sounded like you said your dad's a dancing panda."

"That *is* what I said."

"Then I'm still confused," said Mo.

Will sighed. "Remember when I told you about the guy in the panda suit who saved me from Mark that day?" Mo nodded. "Well, that was my dad."

"Why was he dressed like a panda?"

"He was dancing. In the park."

"Like, for fun?"

"No, like, for money."

"I thought he worked on a building site?"

"He did," said Will, "but he got fired, so he decided to become a dancing panda instead."

"No offense, but, well, I never thought of your dad as the dancing panda type." Mo pondered this for a moment. "Actually, I never thought of *anybody* as the dancing panda type. I didn't even know your dad *could* dance."

"A pole dancer taught him after he rescued her bathrobe that was stolen by a wizard who can set things on fire with his mind," said Will matter-of-factly.

Mo waited for the punch line. None came. "You are totally making this up," he said.

"I *couldn't* make this up."

"For reals?"

"For reals."

"Then this is officially the coolest thing I've ever heard. I think I want to be your dad when I grow up."

"You don't," said Will. "He's a liar and I hate him. And anyway, I thought you wanted to be a zoophile or whatever it's called."

"Screw that. Who wants to be a zoologist when you could be helping pole dancers fight telekinetic wizards? That's the stuff dreams are made of."

"Pole dancers?" said Mark as he swaggered past with Gavin and Tony. "You losers talking about Will's mum again?" He laughed, as did his goons, although even they seemed reluctant to endorse the joke.

"Are you the only person in your family with Tourette's, Mark?" said Mo. "Or did you get it from your mum and dad?"

"What's Tourette's?" whispered Gavin.

"French food," said Tony.

Gavin nodded, even more confused.

"What did you just say?" said Mark, squaring up to Mo, who started to reply before his words were cut short by a hand around his throat. "Don't you *ever* talk about my dad again, you little shit!" spat Mark.

"Why?" said Will. "Because it hurts?"

"What?" said Mark, his grip relaxing on Mo as he turned his sights on Will.

"It hurts, doesn't it?" said Will, his heart racing but his voice steady as he forced himself to hold Mark's gaze. "When people say things about someone you love who isn't here anymore."

"I'll hurt *you* if you don't shut your mouth!"

"Not as much as you're hurting yourself," said Will.

Mark frowned. "What are you talking about?"

"You try to act tough all the time, but I know you feel the same way that I do."

"You don't know shit," said Mark, standing so close to Will that their toes were almost touching.

"I know you keep yourself awake at night wondering why it had to happen to you and not to somebody else," said Will.

"Shut up."

"I know you see people with their mums and dads and you wish that could be you."

"I said shut up!" shouted Mark, his voice cracking slightly.

"I know you hold on to things that belonged to your dad because you think there's still a little piece of him attached to it."

"Shut the fuck up!" yelled Mark, tugging his sleeve over the old silver Casio around his wrist.

"And I know you're angry, Mark," said Will, his voice shaking now. "You're angry because somehow the world just continues, even though your life's been ruined, and it feels so unfair that you want to ruin other people's lives because it's not right that they get to be happy and you don't. And I know you think that nobody else understands what you feel, and most people don't, but I do, Mark."

Will prodded himself in the chest.

"I know how you feel. I know how much it hurts. But hurting other people won't make it hurt any less. It won't make the pain go away. So keep beating me up. Keep taking the piss out of me. Keep pushing me around. It won't change anything, because your dad is gone, just like my mum, and nothing in the world will ever bring them back."

Mark stared at Will with a jaw tight enough to make a crowbar tremble. His chest was heaving and his fists were shaking like two angry dogs on a lead, and Will quietly braced himself for the moment that Mark set them loose; but much to the surprise of Mo, Gavin, Tony, and everybody else who had gathered at a safe distance to watch the altercation unfold, Mark didn't hit Will that day. He didn't even speak. Instead he turned and marched across the schoolyard, his hands no longer balled into fists but shielding his face from the crowd.

CHAPTER 28

Krystal laughed when Danny stood on her foot the first time. She even smiled when he did it the second time. The third time she rolled her eyes, the fourth time she quietly cursed, and the fifth time she cursed so loudly that Fanny stuck her head around the door to check that everything was okay.

"What are you doing?" said Danny as Krystal abandoned him midroutine and angrily turned off the music.

"What am I doing?" said Krystal. "What are you doing, Danny?"

"Er ... dancing?"

"Yeah, all over my fucking feet. These right here are delicate instruments. They're what I use to make money."

"Really?" said Danny doubtfully. "People pay to see your feet?"

"One guy does, actually, smart-arse. And he pays pretty well too, but he won't if my feet look like the cobbles of fucking Pamplona, will he?"

"I told you already, it was an accident. Accidents happen."

"Yeah, you're living proof of that. But five times isn't an accident, Danny. Once is an accident. Twice at a push. But five times? Five times is not an accident. Five times is a fucking joke."

"I said I was sorry."

"What are you sorry for exactly?" said Krystal. "Are you sorry for standing on my feet? Or are you sorry for wasting my time this morning when I could have been in bed watching *Bargain Hunt*?"

"I'm just a little rusty today," said Danny in a tone that failed to convince even himself.

"Rusty? Danny, *watching* you is giving me tetanus. Some of our regulars move better than you and they've had multiple hip replacements. You do realize the competition's in five days, don't you? Five days, Danny, so why the fuck are you dancing like you've got another five *months*? Seriously, if this is all you're going to give, then you may as well just go ahead and break your own legs now, spoil the fun for your landlord at least. I'll give you a hand if you want."

"Look, you're right, and I'm sorry. It's just . . . Will started talking again after all this time, and things have been going really well, but this morning we had this fight and—"

"Don't take this the wrong way, Danny, but I don't give a shit about whatever domestic bollocks you've got going on right now. Not during the hours we're in this room together. And you shouldn't either. All you should care about is winning this fucking competition. You'll have plenty of time to worry about all that other stuff later, but for now, chuck it in the backseat, give it an iPad and a Capri Sun, and focus on the road in front of you. Got that?"

"Got it."

"Great. Now, get in position and dance like your life literally depends on it, and if you step on my feet just one more time, I swear to God I'll stick one of them so far up your arse that you'll be able to tell me what my nail polish tastes like."

Danny kicked off his sodden shoes and flicked on the kettle before hanging his coat on the chair to dry. He was soaked right through to his underpants, but in a weird kind of way he was glad—not so much for the soggy bottom but for the downpour itself, which had started sometime around midday and hadn't let up since. Instead of going to the park after his session with Krystal ended, which he wasn't in the mood to do, and instead of going straight home, which he also didn't want to do because he knew he'd waste the afternoon moping about the fact that he and Will were back to square one, Danny had taken the rain as a sign from the Big Man that he should stay at Fanny's and keep practicing, so that's what he'd done. Only later, when he'd emerged from the club in the late afternoon to find a monsoon where the downpour had been, did he realize that shitty weather and celestial signals were not necessarily the same things.

Will's bedroom door was closed, slammed shut with such force that the nameplate had fallen off and now lay facedown on the carpet. Danny knocked so gently that it almost defied the purpose of knocking.

"Will? You there, mate?"

He placed his ear to the door and thought he heard something, a faint and almost imperceptible sound like the slightest contraction of a bedspring or a sleeve-smothered yawn, but it was hard to determine if the sound came from Will or from the rain on the windows or the kettle in the kitchen. He briefly considered entering uninvited, justifying his intrusion by imagining that Will wasn't in fact ignoring him but simply unable to hear him through the headphones he often wore when playing on his iPad. Then again, it was just as likely that his son was currently burning two holes in the door with the same angry eyes that had burned into him that morning, hearing everything but saying nothing in the way at which he'd become so wearyingly adept. Reluctant to take the gamble, Danny let go of the handle and backed away from the door, telling himself, despite the wealth of evidence to the contrary, that Will would talk when he felt like it.

It was only when he returned with dinner a few hours later that he started to panic a little. His plan had been to lure Will out with his favorite pizza, even placing the box on the floor and fanning the smell beneath the door, but Will still hadn't taken the bait, so Danny decided to bring the bait to him.

"Will, I'm just going to open your door a little bit and leave the pizza inside, okay?" he said, his voice slow and clear like that of a hostage negotiator. "I promise I won't come in. I'd try to slide it underneath, but I asked them to double up on everything so I don't think it'll fit. Is that okay with you? Tell me if it's not okay."

Will didn't respond, so Danny opened the door and pushed the deep-pan Hawaiian inside, nudging it with his fingertips like a novice zookeeper feeding a tiger. He peered into the room, readying himself to slink away at the first sign of a death stare, but what he saw instead unnerved him far more than any hateful expression that his son was capable of mustering, and the boy had quite the repertoire.

Will's bedroom was clean. Not spotless. Not even close to spotless. More dirty than clean actually, but still cleaner than it should have been at that time of day. Will had a tradition that Liz had dubbed "the cleansing ritual," which Danny always found to be a rather philosophical interpretation of their son's habit of scattering his uniform around his bedroom the moment he arrived home from school every day, but there was no limp tie draped over the lamp and no sock perched on the door handle. His schoolbag was also nowhere to be seen.

"Will?" said Danny as he stepped through the door, but even before he opened his mouth he knew he was speaking to an empty room. Will wasn't on his bed. Nor was he at his desk, or under it, or behind the door, or anywhere else for that matter. The only proof that Danny could see that Will had even been home was the nameplate on the floor, and that could have fallen off that morning, for all he knew.

Returning to the living room, he grabbed his phone and checked for any missed calls or messages. Finding none, he dialed Will's number, but the call went straight to voice mail. He tried several more times, each time with the same result.

Guessing he was probably with Mo, Danny called the boy's dad, Yasir, an estate agent with a permanent smile and glasses even thicker than his son's, but the man said Mo was at home watching Animal Planet and hadn't seen Will since school.

"Everything okay?" said Yasir. Lions could be heard devouring something in the background.

Reassuring Yasir that everything was fine and trying to sound confident about it, Danny thanked him and hung up.

"Don't panic," he said to himself, repeating the words like a mantra in the hope that hearing them spoken aloud might help to slow his quickening pulse, but hearing the word *panic* over and over only made matters worse.

He took a deep breath and urged himself to stay calm and think logically. It was barely eight o'clock and it was still light outside, two facts he took comfort in. He also told himself that even though this was massively out of character for Will, his son had left the house that morning angrier than Danny had ever seen him, which meant that he was almost certainly still angry now, which meant he probably didn't want to see the person who had made him angry to begin with, which most likely explained why he hadn't come home yet. Danny couldn't overlook the countless occasions he himself had gone AWOL when he was young—even younger than Will—often as a result of quarrels with his parents, or quarrels *between* his parents. Nothing bad had ever happened to him during those times of self-imposed exile, and he always came home eventually, usually when he was tired, or hungry, or when the fire that burned in his belly was no longer warm enough to keep the chill from his bones.

Reassured by this, Danny took a seat on the couch and waited for the rain to wash Will out from wherever he was hiding and dump him cold and wet on the doorstep. He stared at his phone and listened for the front door, certain that Will would be home any second; but as thirty minutes passed, and then doubled, Danny became increasingly restless, especially as the sky grew dark. Unable to wait any longer without drumming a hole through the floor with his foot or shredding the arms of the couch with his nails, he grabbed his still-wet jacket from the chair and ran back out into the pouring rain.

He aimed for the wooden house in the children's play area where Mo and Will liked to sit sometimes after school, despite having to fold themselves almost in half to fit inside, but all he found were some empty nitrous oxide canisters and the damp remains of a Happy Meal. He checked the row of garages behind their flat, some of which had been prised open with crowbars by sinewy teenagers who used them as hangout spots, hookup dens, and makeshift offices for various, mostly illegal enterprises, but Will wasn't there either. Remembering that kids sometimes liked to root around in the pile of scrap near the bins where televisions, rotting furniture, and other household items were often dumped by tenants or land-lords, he went there next, his heart pounding through his jacket and his feet slapping through the muddy puddles, but all he found were a couple of cats hiding from the rain beneath a three-legged table.

Danny ran up and down the apartment stairwells and along every corridor of every level of the building until his thighs ached and his lungs burned and his throat was sore from shouting Will's name and yelling at people who were yelling at him for shouting. He wanted to scour every corner of the city, to navigate every dark alley and every busy road, to search every shady park and every neon underpass, but he didn't know where to begin and knew he had more chance of counting the raindrops than he did of finding

Will by blindly walking the streets. He felt as helpless as he had when his son was lying comatose in the hospital and Liz was in the morgue; and the worst part was knowing that no matter how much he appealed to a God he didn't believe in, and no matter how much he tried to convince himself that the world was a fair and just place that functioned according to logic and not according to chance, nothing he said, nothing he did, and nothing he quietly muttered in prayer could change the course of what was unfolding.

Pausing to catch his breath, he gripped the railing and stared through the rain towards the hazy skyline of Central London. Since the moment he'd discovered that Will was missing, Danny had guessed that their fight was to blame, but the longer he stared at the dark mass of buildings that loomed on the horizon, the more his mind drifted to the darker corners of his imagination. He thought about all the bad things that shouldn't happen to good people but did, and all the bad people who should obey the law but didn't. He recalled all the ugly mug shots and grisly headlines he'd seen in the newspapers over the years, and all the grim reports he'd heard from dour newsreaders and TV presenters urging people to come forward with information. He pictured all the missing-person posters taped to lampposts and walls and bins and electricity boxes that he'd passed in his life without so much as a glance, and he imagined Will's face on one of them, his image torn and weather-beaten, flapping in the wind and ignored like the others.

As the excruciating realization dawned on him that Will might not have disappeared by choice, he reached into his jacket, ready to call the police, a call he hadn't wanted to believe was necessary until then; but patting his pockets and finding them empty, Danny realized he'd left his phone at home.

Racing along the corridor and almost losing his footing as he barreled down the stairwell four steps at a time, he reached the front door and fumbled with his keys, cursing himself for every wasted second as he dropped them twice and repeatedly missed the lock until, steadying his trembling hand with the other, he finally guided the key into place.

Hoping to find a box full of pizza crusts, Danny returned to an empty flat. No sopping size sixes in the hallway. No schoolbag slumped in the corner. No uniform tossed around the carpet. No angry eleven-year-old waiting to ignore him. He grabbed his phone from the table and made the call that every parent dreaded.

"What service do you require?" said the female operator.

"Police," said Danny.

"One moment," she said, putting him on hold.

Danny flicked through the transcripts of his conversations with Will while he waited, hoping to find some previously overlooked clue that might give him an indication of his son's whereabouts, some passing reference to a friend he'd never heard of perhaps, or a hangout spot he didn't know about; but the notes only served to remind Danny of just how close he and Will had been recently, and how far apart the two of them were now.

"What's your emergency?" said the operator, this time male.

"My son is missing," said Danny, barely able to believe the words coming out of his mouth.

The operator began to ask various questions about Will. Name. Age. Height. Date of birth. What he was wearing when he disappeared. When he was last seen. Where he was last seen. Danny felt numb as he answered everything that was put to him, as if he weren't actually talking but listening to somebody else answering the questions for him while he stared at the notepad in his hand. In the corner of the page, in rough scribble, he saw he'd written the word *oranges*. He'd underlined it twice and put a question mark at the end of it, but he struggled to understand the significance of the word until he suddenly remembered what Will had told him that day in the park. He dared to smile, but only for a second.

"I'm sorry," he said, cutting the operator off in the middle of a question as he hurried towards Will's room. "It's fine. Everything's fine. Sorry. Sorry for wasting your time."

Danny ended the call and stopped in front of the wardrobe, the one place he hadn't thought to check. He gently slid the door aside, and the unmistakable scent of Liz drifted out and embraced him for just long enough to let him know that everything was going to be okay. Will was curled up in the corner with his head resting on his backpack. He still wore his shoes and uniform, and his headphones were over his ears. He didn't move when Danny found him. He didn't even look at him, too deeply asleep to realize he'd been found or to understand he'd been missing in the first place. Next to his foot was a small plastic container with an orange lid beside it, the type Liz had kept on her bedside table along with her phone and whatever book she happened to be reading at the time. The jar was empty save for a small dab of hand cream at the bottom, just about the amount she used first thing in the morning or last thing at night.

Danny screwed the lid back on and thought about waking Will, but noticing how peaceful he looked, he quietly closed the wardrobe door and tiptoed out of the room.

CHAPTER 29

Danny had never been insulted by a pigeon before. He'd never been insulted by any kind of animal as far as he could recall, but as he squinted against the sunlight that streamed through the open curtains and looked at the pigeon that was currently eyeballing him from the ledge outside his window, cocking its head from left to right as if contemplating one or all of life's biggest conundrums, Danny felt strangely confident that this bird, despite its seemingly innocent appearance and its biological inability to form words, had nevertheless just called him a fuckbag. Only when he heard Krystal yelling through the letterbox did he understand where the insults were coming from.

"Open up, you fucking pillock!" she shouted. Danny could hear the door shaking, perhaps from fear or perhaps from the fist she was hammering against it. "I know you're in there!"

Snatching the clock from the bedside table, Danny cursed when he saw the time. Realizing he'd forgotten to set an alarm and was consequently two hours late for dance practice, he leapt out of bed, threw on some clothes, scurried down the hallway, and opened the door.

"I can explain," he said. Krystal's fist was suspended in the air and he flinched, unsure if her knuckles were destined for the door or his nose.

"You've got ten seconds, after which time I'm going to mace you," said Krystal. She pulled a tiny canister from her bag and aimed it at the gap between Danny's eyebrows. "I've got to warn you, though, unless you have a fucking good excuse, and by 'good' I mean, I don't know, you were arrested for pissing in a policeman's helmet, or you were kidnapped by sex traffickers and then returned because nobody wanted to buy you, then there's a very high chance that you're going to be spending the rest of the morning pouring milk in your eyes."

"Why milk?"

"It helps to stop the burn."

"Got it," said Danny. "Can we do this inside though? We've got neighbors. And, well, the milk's in the fridge."

Krystal thought about this for a moment. "Okay, move it," she said.

Danny backed down the hallway. Krystal followed him into the kitchen.

"One thing," said Danny. "Does it matter if it's full-fat or semi-skimmed? Because I've only got semi—"

"Ten seconds."

"Okay, okay. Will went missing last night. I couldn't find him anywhere, I called the police and everything. I didn't get home until late and, well, I guess I forgot to set my alarm. There."

Danny waited for Krystal to holster her weapon. She didn't.

"Five," she said.

Danny frowned. "Five what?"

"Four."

"I just told you everything!"

"Three."

"Wait!"

"Two."

"Can you at least answer my question about the milk?"

"One."

"I was kidnapped by sex traffickers!"

"Nice try," said Krystal. She mashed the nozzle and Danny screamed as the jet hit him right between the eyes. He clawed at his face while she casually retrieved a carton from the fridge.

"Here," she said, handing him the milk. Danny grabbed it, and Krystal laughed as she watched him pour it over his head.

"Was that really necessary?" said Danny as he wiped his face with a tea towel.

"Spraying you with Silly String, you mean? Or letting you pour milk all over your head?"

"Silly String?!" said Danny, only then noticing the matted mess of colored threads stuck between his fingers.

"You didn't really think I'd waste a decent can of Mace on you, did you?" she said, returning the canister to her bag.

"I'm not sure if I'm supposed to be flattered by that or not."

"Seriously, you should have seen your face. It was literally the most pathetic thing I've ever seen. I wish I'd filmed it, that shit would have gone viral for sure. Actually, can we do it again?"

"We're out of milk."

"Shame, I was gagging for a brew."

"Forgive me if I don't give you any sympathy."

"An apology will do," she said.

Danny laughed. "An apology?" he said. "For what?"

"Hmm. That's a tough one. Let me think for a second. Oh yeah! That's right, for fucking leaving me hanging around Fanny's on a fucking Saturday morning, otherwise known as my fucking day off, that's fucking what."

"Yeah, well, I'd say we're pretty fucking even," said Danny as he wrung the milk from his T-shirt. "And anyway, I told you why I couldn't make it. If that's not a good enough reason, then I don't know what is."

"I thought you just made that stuff up so I wouldn't mace you," said Krystal.

"I *wish* I was making it up," said Danny. He grabbed a few bowls and a box of Coco Pops and carried them to the table.

"What happened?" she said, taking a seat.

Danny sat opposite and explained everything: the fight in the morning, Will going missing, running around in the rain, calling the police, and finally discovering where he was hiding.

"Wait, so he was here all along?" said Krystal.

"Yep. Fast asleep in the wardrobe."

"That is hi-lar-i-ous," she said, turning one word into four. "I mean, it's terrible, obviously, but it's also kind of funny, right?"

"No," said Danny. "It's not."

"Just a tiny little bit?" She measured a tiny little bit with her thumb and forefinger and peered at him through the gap. Danny stared at her. Krystal sighed.

"All right, serious Simon, whatever," she said. "So where is he now? In the cupboard? Under the table?"

"Right here," said Danny as Will shuffled into the living room with the hairstyle of somebody who had just spent the night in a wardrobe. "Morning, mate."

Will said nothing as he stared at Krystal.

"Hello, trouble," she said. "You must be Will."

Will nodded, trying to look at her without actually looking at her like he sometimes did with Victoria's Secret window displays.

"You never said he was so good-looking," Krystal said to Danny. "Look at his eyes, they're bluer than a bishop's balls."

Danny frowned at her.

"What? They are! You sure he's yours?"

"He takes after his mum," said Danny, nodding towards the framed photograph of Liz.

"Lucky for you," she said to Will. "Hope you got her brains too."

"Very funny. Will, this is my friend Krystal."

"Pleasure's all mine," she said as Will bashfully shook her hand. "I'm not actually his friend, by the way. Your dad doesn't have any friends. I bet you do, though, don't you? I bet all the girls want to be your friend."

Will laughed nervously. He somehow managed to shrug, nod, and shake his head all at the same time.

"Sleep okay?" said Danny, attempting to change the subject before his son turned any redder.

Will ignored him and reached for the Coco Pops. "Can you ask my dad to pass the milk please?" he said to Krystal.

"You got a mouth, ain't you?"

"I'm not talking to him."

"Why?"

"Because he's a liar."

"He's a thief too. You know he stole money from me?"

"I didn't *steal* it," said Danny. "I said I'd pay you back."

"Whatever. Milk."

Danny passed the carton to Krystal, who passed it to Will, who upended it over his bowl and watched a sad trickle of milk dribble onto his cereal.

"Can you ask my dad why there's no milk?" he said.

Danny looked at Krystal. "I think you can answer that one," he said.

"He poured it over his head."

"Why?"

"Because he's an idiot."

"Krystal tried to mace me," said Danny.

"Did you really?" said Will with more enthusiasm than Danny was comfortable with.

"Sort of," she said.

"Cool," said Will, digging into his arid cereal. "He probably deserved it."

"I like him," said Krystal, nodding at Will.

Danny rolled his eyes.

"How do you know my dad anyway?" said Will through a mouthful of dry Coco Pops.

"God hates me, so he sent Danny as punishment," said Krystal.

"Because you're a stripper?"

"Who said I was a stripper?"

Will pointed at Danny. Danny squirmed.

"The *panda* told him you were a stripper, actually," he said. "Blame him."

"Well, the *panda's* talking out of his furry arsehole, because I'm not a stripper. I'm a pole dancer, Will, and there's a massive difference. See, anybody can be a stripper, it's easy. All you gotta do is take off your clothes and wiggle your bits in somebody's face—"

"Yeah, see, I don't think Will needs to know about—"

"—But pole dancing? That's a different thing altogether. Pole dancing is a skill. It takes a lot of hard work and practice. Pole dancers aren't just dancers. They're *artists,* mate. Us lot, we're like the Leonardo da Vincis of the entertainment industry, but we're better, because even he couldn't do what we do."

"Why?" said Will.

Krystal shrugged. "Not flexible enough."

"And on that note, we should probably get going," said Danny, keen to bring the conversation to an end.

"Going where?" said Krystal.

"The club. To practice."

"No chance, mate. I've got an appointment for a Hollywood at noon."

"What's a Hollywood?" said Will.

"It's like a Brazilian, but more painful."

"What's a Brazilian?"

"Ask your dad," said Krystal.

"I'm not talking to him."

"Precisely."

"Can't it wait?" said Danny. "The competition's in four days!"

"Yeah, which is why I was ready at eight o'clock this morning while you were still tucked up in your jimjams."

"I told you already, I was looking for Will!"

"And my regulars'll be looking for new places to stuff their money if I miss this appointment, and appointments with Fernando are *not* easy to come by. He's amazing. He's like the Mr. Miyagi of the waxing world."

"You were looking for me?" said Will, his stonewall crumbling beneath the weight of his curiosity. This time it was Danny's turn to remain silent.

"Are you going to tell him or am I?" said Krystal.

"Tell me what?" said Will.

Danny slumped in his chair like a boxer whose trainer had just thrown in the towel. "I thought you'd run away," he said. "I came home yesterday and you weren't here and, well, I thought you'd taken off because you were still angry with me about the whole panda thing."

"He called the police and everything," said Krystal.

"Is that true?" said Will.

Danny nodded. "I didn't know what else to do. Losing your mum was bad enough, but the thought of losing you . . ." His hand trembled as he absently circled a rogue Coco Pop with his finger. Krystal put her hand on his shoulder and gave it a gentle squeeze. "You're my best friend, Will. I know you don't believe me, but it's true. I honestly don't know what I'd do without you, mate."

"Then why didn't you tell me about the panda thing?"

"Because I was embarrassed, that's why. I thought you'd be ashamed of me. *I'd* be ashamed of me. Who wants a dancing panda for a dad?"

"Me!" said Will, prodding himself in the chest. "I'm not angry about you being a panda, Dad. I'm angry that you didn't tell me about it. I'm proud of what you're doing. I saw you in the park, you're awesome. I never knew you could dance like that."

"He couldn't until he met me," said Krystal. "Just saying."

"Thanks, mate," said Danny. "Let's hope the judges feel the same."

"Judges?" said Will.

"Battle of the Street Performers," said Krystal. "It's like *X Factor* for homeless people."

"But Dad's not homeless," said Will.

Danny sighed. "Not yet, I'm not."

"What do you mean?"

"I mean that we *really* need to win this competition," said Danny, wanting to be honest without being *too* honest.

Will looked at Danny and Krystal as he took a moment to process this.

"So win," he said matter-of-factly, as if winning required nothing more than a verbal commitment to do so. He closed the matter with a mouthful of cereal.

"I can't. Not without Krystal."

"Nice try," she said without looking up from her phone, "but I am officially immune to emotional blackmail. Five years on a pole will do that to a girl."

Will fixed Krystal with the same pleading eyes that he used to give Liz when the ice cream truck was approaching.

"Don't look at me like that," she said. Will didn't blink. "Danny, will you please tell your son to stop looking at me like that?"

"He'll stop if you agree to help us."

"I told you already, I've got plans."

"Please," said Will.

"No."

"Pretty please."

"No!"

"Extra please."

"Extra no."

"I'll give you a hundred pounds," said Will.

Krystal burst out laughing, her air of indifference suddenly obliterated.

"I thought you said he takes after his mum?" she said. Danny held his hands up like somebody caught near a shop display that had just collapsed. "You really are a chip off the old block," she said to Will.

"Thanks," he said.

"It wasn't a compliment," said Krystal.

"Please," said Danny. "Will you help us?"

Krystal stared at Danny, and then at Will, and then at the space between them both.

She sighed. "What did I ever do to get lumbered with you two?"

"Is that a yes?" said Danny.

"It's a hurry the f—" Krystal glanced at Will. "—hurry the frig up before I change my frigging mind," she said. Danny and Will shared a secret low-five beneath the table as she stood and adjusted her miniskirt. "Might want to change out of that shirt first though," she said, pointing at Danny. "And you might want to get out of those pj's," she said, this time looking at Will.

"Where am *I* going?" said Will.

"She's just joking," said Danny.

"Do I look like I'm joking?" said Krystal. "If I'm giving up my Saturday, then so are you. Come on, chop-chop!"

CHAPTER 30

Danny lingered outside the club as Krystal disappeared inside.

"Promise me you won't tell anyone about this," he said, looking up and down the street like a nervous getaway driver.

"I promise," said Will as the doors clattered open and two stocky men emerged carrying one of the black-and-white cow-skin couches from the VIP lounge.

"Good," said Danny, watching the men dump the furniture in a nearby alley full of dumpsters and plastic beer crates. "And promise you'll keep your eyes closed until I tell you to open them. There are things in there that you shouldn't be looking at. Not for another few years at least."

"Like boobs?" said Will.

Danny sighed. "Just keep your eyes closed, okay?"

Will rolled his eyes and covered them.

"No peeking," said Danny as he guided him into the club.

An old lady in a blue tabard was busy lugging an even older vacuum cleaner around the well-worn carpet, which wasn't so much in need of a clean as a Viking sea burial. Sitting side by side on the podium, two women dressed in nothing but bikini bottoms were casually passing a cigarette back and forth while they waited for the first of the punters to arrive. They both waved at Danny, who sheepishly waved back. So did Will, essentially blowing his cover in the process.

"I said no peeking!" said Danny as he briskly ushered Will through the door behind the bar.

They found Krystal on the floor, resting her forehead on her outstretched legs as she warmed up for the second time that morning.

"Is Fanny renovating or something?" said Danny.

"What?" she said, looking up.

"The couches," said Danny. He jerked his thumb towards the main room.

"Oh. Yeah. The health inspectors are coming next week, and Fanny was scared they might be contaminated."

"With what?"

"With more DNA than the national crime database," she said. Danny shuddered. "I told her I knew a bloke who could fumigate them for her, but she didn't want to take any chances."

"At least one Fanny's getting a refurb," said Danny. He briefly chuckled at his own joke before he saw how close he was to being murdered.

"Mum would have loved this," said Will, running his fingers across the mirrors and pressing the creaky floorboard with his toes. "All the dancing space, I mean. Not the club. I don't think she'd like the club. No offense," he said to Krystal.

"None taken," she said, climbing to her feet. "I don't like it much either, which is why I really, *really* cherish my days off." She scowled at Danny, who pretended not to notice as he focused instead on his own series of less demanding warm-ups.

"Okay," said Krystal once Danny had sufficiently limbered up. "Get that panda suit on and show us what you've got. I want to see the whole routine from start to finish."

She unstacked a couple of plastic chairs and placed them side by side.

"Will," she said, sitting down and gesturing to the chair beside her. "Come and join me on the judging panel." Will took a seat. "I want you to watch everything that furry fool does, and tell me if something doesn't look right. Got it?"

"Got it," said Will. He shuffled to the edge of his seat.

"Good. And remember to be mean. I know he's your dad, but you can't go easy on him. You've got to act like a judge, and judges are arseholes, okay? Now, show us your mean face."

Will pictured Reg's face and tried to emulate his permanent scowl.

"I said mean, not constipated! Try again, like this."

Krystal pulled a face that made Danny's fur stand on end. Will did his best to follow suit.

"Much better," she said. "Danny, over to you."

The music kicked in and Danny got to work, his eyes fixed firmly on the mirror. He'd never performed the entire act in front of Krystal before, and doing it now made him nervous enough, but performing in front of her *and* Will made his sweat glands open like the starting gates at a greyhound track. Still, aside from stumbling a couple of times and missing a handful of beats, blunders that he felt he recovered from well with various spontaneous adjustments, Danny made it through the performance soggy and spent but mostly unscathed.

"So?" he said, pulling his mask off and wiping his face. Krystal gestured for Will to go first.

"I thought it was terrible," he said.

"What!" said Danny.

Even Krystal look shocked. "Seriously?" she said.

"No, but you told me to be mean."

"Yeah, not *that* mean."

"Oh."

"What did you really think?" she said.

"I thought it was awesome!" said Will. Danny shot him a thumbs-up.

"Okay," said Krystal. She turned to Danny. "Well, the good news is that it wasn't terrible, but the bad news is that it wasn't even close to awesome. You won't come last, but you sure as buggery won't come first either, not unless you drop the improvs. I don't know if they were deliberate or whether you'd just forgotten the routine—you know, the one we've been through *a bajillion times*—but stick to the script next time."

"The judges don't know the script, so how are they going to know if I stick to it or not?" said Danny a little defensively.

"Balance, Danny. The routine is supposed to have balance. Right now you've got too many spins in one section and too much fancy footwork in the other. The judges will either think you're improvising or that your choreography is crap. Either way, they're going to mark you down for it. Also, you're still screwing up four or five moves. That kick-twist needs more work, for a start."

She stood and positioned herself beside Danny. Will followed, keen to observe the process.

"See, it needs to be more like this," she said, effortlessly pulling off the move in question.

"Like this?" said Danny, trying to mimic her movements.

"No, see, you're still dragging the foot," she said, performing the move again, but slower this time. Danny tried a second time with similar results.

"Like this?" said Will, kicking and twisting just as Krystal had instructed.

"Yes!" she said. "Exactly! Do that again."

Will pulled off another perfect kick-twist.

"You, mate, are a natural. Any chance you can show your old man how to do that?"

"Put your foot like this," said Will, instructing Danny to follow his lead. "Then kick off like this, and then when you spin you need to land in this position right here."

Danny watched him again before going through the motions himself.

"Perfect!" said Krystal, low-fiving Will and high-fiving Danny in the side of the head. "That wasn't so hard, was it?"

"I guess not," said Danny, surprised by how easy it was.

"And that other move you did, the one that went like this," said Will, spinning and dropping to his knees before returning to his feet in one fluid motion. "You need to twist faster when you get up, otherwise you'll keep missing the next beat, right, Krystal?"

"Right," she said. She looked at Danny and gave him her best what-the-fuck look. "Exactly. Do that again."

Will repeated the move, emphasizing the twist so Danny could see it better.

"Like this?" said Danny, but even as he spoke he knew that he'd nailed it.

"Yeah," said Will. "Then you can move into that next section." He pulled off the next few moves in the sequence while Krystal watched in amazement.

"Have you seen this routine before?" she said.

"No," said Will.

"Then how do you know so much of it?"

"Because I just watched it."

"And you've already memorized the whole thing?"

"No," said Will. "Not the whole thing. But most of it probably."

"Danny, get your arse over here," she said, sitting down and patting the chair beside her. Danny did as he was told. "Will, can I ask you a big favor?"

"I guess," he said, a little unsurely.

"Would you mind pressing play on that stereo and showing us everything you can remember?"

Will looked at the stereo and shrugged.

"Okay," he said.

Danny barely breathed and Krystal hardly blinked for the next three minutes and twelve seconds, their vital functions held hostage by Will as he tore his way through their performance. He couldn't remember every single move, but he hit far more than he missed, especially the more challenging maneuvers that Danny had struggled with the most; and when the music stopped almost two hundred dance moves later, the only sound that could be heard in the otherwise silent studio was the rasp of Will sucking air into his lungs as Danny and Krystal stared at each other in disbelief.

"How the flipping flip do you flipping know how to dance like that?" said Krystal, her bracelets jangling like wind chimes as she pounded her palms in applause.

"His mum," said Danny. He smiled at Will, who nodded in agreement.

"You mean to tell me you've had a perfectly good dance teacher living under your roof all this bloody time and you didn't think that information was maybe worth sharing?"

"I didn't know," he said, the pride in his voice partly hiding the shame.

"You realize how important this is, don't you?" said Krystal. She leapt out of her chair. "You understand what this means, right?"

"No," said Danny, struggling to follow. "What does this mean?"

"Jesus, do I really have to spell it out for you? It means, *Danny*, that I don't have to cancel my appointment with Fernando!" She grabbed her coat and made for the door. "I'll be back in a few hours. Will, teach him everything he needs to know."

"Wait!" yelped Danny, but Krystal had already gone.

He stared at the door as it creaked to a halt, sure she'd be back any minute. Then, realizing she was serious, he turned to Will, who was still standing awkwardly in the middle of the room, and smoothed his fur with his palms.

"I guess we better get started," he said.

*   *   *

Apart from a suspiciously long bathroom break for Will, which only ended when Vesuvius escorted him back to the studio after Will had "accidentally" taken a wrong turn and "accidentally" ended up in the VIP lounge, the two of them didn't leave the room for the next few hours. Nor did they stop dancing, working their way through the routine over and again while the flimsy walls and loose floor-boards shook in time to the song that throbbed from the speakers on repeat. They danced side by side, watching each other's footwork in the mirror in front of them, Danny looking for guidance and Will looking for faults. Every time Danny fluffed a move or accidentally deviated from the routine, Will would stop the music and walk him through the troublesome step or section before restarting the track and going through the performance from the beginning. If neither of them was sure of a particular move or sequence, they would huddle around Danny's phone to watch a recording that Danny had taken of Krystal demonstrating the routine from start to finish, which he often consulted during his solo practice sessions. Unlike when Danny would try and often fail to replicate Krystal's moves directly from the video, this time he watched Will perform them before attempting his son's renditions.

The approach was something like a dancing version of a game of Telephone, but instead of the message getting lost or scrambled as it so often did in the children's game, here the process had the opposite effect, taking something complex and making it simple without losing any of its value. Krystal had taught Danny a lot since she'd reluctantly agreed to help him, but her teaching skills often left as much to be desired as his ability to learn. She wasn't always able to deconstruct her experience in a way that Danny could digest, which led to frustration on both sides as Krystal grew impatient and Danny grew more flustered and more prone to making mistakes.

Will, on the other hand, was able to explain everything to him like an eleven-year-old. Because of this, and combined with the skills and knowledge that Danny himself had accrued over the last couple of months, the two of them were able to iron out almost every crease by the time Krystal returned a few hours later. She watched them undetected from the corner of the room while they worked their way through the routine, and only when the music stopped and she started applauding did they realize they weren't alone.

"That, fellas, was smoother than my lady bits," she said. "Looks like your new teacher here has been doing a much better job than I have." She winked at Will, who smiled shyly.

"What do you think?" said Danny.

"What do I think?" she said. "I think I've been wasting my bloody time this last couple of months, Danny, that's what I think. You never needed me. All you needed was this little mover right here." She gently nudged Will in the ribs. "I guess we're going to have to split that prize money three ways now. I still get half though, obviously."

"I hope it'll be enough," said Danny.

Krystal shrugged. "You can give me more if you like," she said.

Danny rolled his eyes. "I meant the performance," he said. "Do you think it's enough to convince the judges?"

"If it isn't, I really don't know what will be," said Krystal. "Not unless . . ." She trailed off, suddenly lost in thought.

"Unless what?" said Danny.

Krystal nibbled her bottom lip and stared into the middle distance. "Where are those couches?" she said.

"What?" said Danny.

"The ones that Fanny chucked out, where are they?"

"In the alley opposite the club, why?"

"You'll see," she said, heading for the door. "Follow me. I have a brilliant idea."

CHAPTER 31

Danny was a little disappointed when he first set eyes on the stage. He'd spent the last month imagining some kind of setup akin to that of a U2 concert, with huge television screens and speakers the size of small apartment buildings; but when he visited Hyde Park the day before the competition to get a feel for what to expect, the venue he encountered looked more suited to a Punch-and-Judy show than the Battle of the Street Performers so boldly advertised on the flyer he'd carried around in his wallet for the last few weeks. The only battle that Danny could foresee taking place was the battle he was going to have trying to perform his routine without falling off the laughably small stage that the crew were still in the process of building. It wasn't just the stage that hadn't been finished either—at least, that's what Danny hoped, because if everything else had been completed, the audience would have a few battles of their own, like how and where to go to the bathroom when there were no toilets, or how they were going to see the stage when there were no lighting rigs to speak of, or how they were going to find the event in the first place when there were no banners, posters, or advertisements.

Wondering if he'd somehow missed the competition and if the stage construction he was currently witnessing was in fact *de*construction, he fished the flyer from his pocket to reconfirm the date of the event. Seeing that it was indeed scheduled for tomorrow, he folded up the flyer and frowned as he watched the crew taking yet another cigarette break.

Danny's disappointment began to subside when he thought about it later that evening. The size of the venue didn't matter. The only thing that mattered was the size of the prize. He'd happily perform in a supermarket car park just as long as the ten grand was still up for grabs. Also, as much as he liked to imagine himself rocking out to the deafening chorus of thousands of screaming admirers, he wasn't even confident about performing his act in front of Krystal and Will, never mind a huge crowd of strangers. The smaller the venue, the smaller the crowd, and fewer people equaled fewer things to worry about, which meant his mind would be free to worry about all the other stuff instead, like the grim repercussions of losing the competition, for example, or the angry rash on his inner thigh that wouldn't go away no matter how much talc he applied, or the actual level of brilliance of Krystal's "brilliant idea." Danny wasn't convinced that it was even a mildly *good* idea, let alone a brilliant one, but with only one day to go before the competition and with the plan already fully integrated into his routine, it was too late to change anything now. And anyway, he reminded himself, at least there wouldn't be many people to see him screw up if things didn't go according to plan. He took some comfort in that thought—or he did until he, Krystal, and Will arrived at Hyde Park the following night to find thousands of people gathered around a stage much bigger and considerably more intimidating than the one he'd watched being lazily constructed just twenty-four hours ago.

"I thought you said it was tiny?" said Krystal, nodding towards the huge rectangular platform that rose from the crowd.

"It was!" said Danny as he looked at the massive lighting rigs that now loomed ominously over the stage.

"This place is awesome!" said Will, his eyes flickering from the strobes and halogens that flashed and panned from the trusses and towers. And Danny had to agree. It *was* awesome, almost terrifyingly so. In fact, it was so awesome that he wondered if perhaps another event was taking place in the park simultaneously, an event more worthy of the beer tents and food trucks and the various television crews that were trying to report their respective pieces to camera, while handfuls of excitable revelers pulled faces and made masturbatory hand gestures behind them. It was only when he neared the stage and saw the huge BATTLE OF THE STREET PERFORMERS banner draped over the front of it that Danny realized, with a swell of pride and a tsunami of fear, that this was indeed the right venue.

"This way!" shouted Krystal. She pointed towards a large tented area enclosed by a fence and guarded by several men who looked like they'd killed the original guards and stolen their uniforms to avoid being caught and returned to whatever prison they'd recently broken out of.

"ID," grunted one of the men blocking the entrance, who appeared to be composed largely of biceps. His biceps had biceps. His triceps had biceps. Even his head looked like a bicep. Danny handed over his street performer's license.

"Name?" said the man, turning the card over in his hand.

Danny frowned. "It's right there on the card," he said, tapping the license.

"Your *stage* name," said the man wearily. "What's your performance called?"

"My performance?" Danny looked at Krystal. Krystal shrugged. "God knows what," he said.

"Nope, sorry," said the man.

"Sorry?"

"You can't have that."

"I . . . can't have what?" said Danny.

"God Knows What."

"What?"

"You can't have God Knows What," said the man as he scanned the clipboard in his hand.

"I can't have God knows what?"

"That's what I just said."

"Wait, so you don't know what I can't have?"

"What?"

"If I can't have God knows what, then what you're basically saying is that you don't know what I can't have."

The man stared at Danny like he was a crossword puzzle on the back of a cereal box.

"What are you on about?" he said.

"Me? What are *you* on about!" said Danny, unaware that a small queue was forming behind him.

"Christ almighty," said the man, the biceps on his biceps beginning to twitch. "Listen to me very carefully because I'm not going to say this again. You can't call your *act* God Knows What. We've already got a religious rock band called that, got it? So you're going to have to think of a different name."

"Pandamonium!" shouted Will. Everybody looked at him. "Get it? Panda? Monium? Pandamonium?"

"That's not half bad, actually," said Krystal.

"Pandamonium it is, then," said Danny.

"Whatever," said the man. He scribbled the word down and grabbed a rubber stamp, which he brandished like a murder weapon. Danny reluctantly extended his arm, and the man mashed the back of his hand with such force that the letters VIP would have been visible even without the ink. After stamping the hands of Krystal and Will with a tenderness he hadn't shown Danny, he stepped aside and allowed them into the performers' area.

"Cubicle twenty-seven!" he shouted as they slowly made their way down the long corridor that ran through the center of the tent. There were countless small canvas partitions identified by numbers above the doors. A few of them were zipped shut, but most were open to reveal various performers in various stages of rehearsal. Some Danny recognized from the park, like the nut juggler, the chicken man, and the human statue, who might have been practicing or who might have simply been sitting very still. Tim was also there, strumming his guitar and twiddling the tuning pegs while Milton sat on his shoulder in a fetching lime-green V-neck sweater; but for every familiar face he passed there were countless other people that Danny had never seen. There were jugglers, there were clowns, there were unicyclists. There was a juggling clown on a unicycle. One cubicle contained a skinny old man wearing a white T-shirt with a Jack Russell's face on it, the same Jack Russell that was sitting on a chair opposite and barking every time the man paused his toothless rendition (in both senses of the word) of "Wannabe" by the Spice Girls.

Another man sporting a tuxedo at least three sizes too small for him was standing behind an upturned hat situated in the middle of a table.

"Ladies and gentlemen," he said to his imaginary audience while he wiggled his fingers mysteriously, "I will now pull a rabbit out of this hat!"

He plunged his hand into the top hat and fished around for a moment before delving deeper, first up to his elbow and then up to his shoulder. Pulling his arm out, he crouched down, peeked beneath the tablecloth, disappeared under the table completely, and then emerged a minute later looking flustered and slightly disheveled.

"Shit," he said, tugging a red handkerchief from his breast pocket and mopping his brow, unaware that several other handkerchiefs had also been dislodged and were now dangling from his pocket like a string of Tibetan prayer flags.

"Here we are," said Krystal, pausing outside number twenty-seven. The canvas walls fluttered as she unzipped the door and entered the tiny cubicle.

"Well, this is cozy," said Danny, sitting on the foldable chair in the corner, which took up about half of the floor space. "I can barely stretch in here, never mind practice."

Ivan's head appeared around the door.

"Room for one more?" he said, forcing the others into various corners as he joined them in the cubicle.

"Ivan!" said Will.

"Hey, Ivan," said Danny. "How did you get in here?"

Ivan turned around to reveal the word *CREW* written across the back of his black T-shirt.

"Did you just murder somebody? Tell me honestly."

"No murder," said Ivan. "eBay. I buy long time ago. Is useful. One time I get into Michael Bolton concert for free because of this T-shirt."

"You snuck into a Michael Bolton concert?" said Krystal, looking at him like he'd just attempted a drunken backflip and failed.

"As a test," said Ivan, trying to sound nonchalant. "You know. To test T-shirt."

"Ivan, this is Krystal, my dance teacher. Krystal, Ivan."

Ivan shook her hand, which disappeared in his.

"My dad saved his life once," said Will, nodding at Ivan. "Didn't you, Dad?"

Krystal and Ivan stared at Danny. Krystal looked doubtful. Ivan looked dangerous.

"What?" said Danny with a nervous laugh when he saw Ivan's expression. "Liz told him that, not me!"

"And who told Liz?" said Ivan.

Before he could answer, Mo appeared in the doorway, much to Danny's relief.

"How did you get past security?" said Danny, pressing himself against the wall as Mo insisted on entering the cubicle.

"I said I had special needs," he said, tapping his hearing aid. "Works every time."

"That's not a lie," said Will. "You *do* have special needs." Mo thumped him in the arm.

"Malooley?" called somebody from the corridor. A second later a man arrived wearing a T-shirt like Ivan's.

"That's me," said Danny.

"And me," said Will.

"And him," said Danny.

"Congratulations," said the man, consulting his clipboard. "You are officially the last act of the evening. I'll give you a shout when you're up."

"Last!" said Danny once the man had gone.

"Last isn't so bad," said Krystal. "I mean, yeah, sure, you've got to sit here and wait until the end of the show, getting more and more nervous while your confidence slowly trickles away until you're a complete emotional wreck. So from *that* perspective, then yeah, it's bad."

"Is this supposed to be motivating?" said Danny.

"I ain't done. Last also gives you an advantage. See, the judges are going to start forgetting all them other acts the moment they're finished, right? But you, you're the last thing they're going to see before they make their final decision. You'll still be fresh in their minds."

"And if you screw up, then you will also be fresh in their minds," said Ivan unhelpfully.

"Thanks, Ivan," said Danny and Krystal together.

"Good luck, Mr. Malooley," said Mo. "You're going to rock!" He made two sets of devil's horns and wiggled them at Danny.

"What's that noise?" said Krystal. Everybody went quiet and listened as somebody spoke into a microphone outside.

"It's the show," said Will. "It's starting!"

CHAPTER 32

"Good evening, Hyde Park!" said a man in his sixties who hobbled onto the stage to the sound of tepid applause. His face was almost as creased as his suit, and he mopped at his brow with a handkerchief.

"Are we all having fun?" He held the microphone over the audience.

Everyone mumbled in stale acknowledgment. Somebody shouted, "Wanker!" and a few people laughed, but the host shrugged it off like a man whose entire life had been spent being called a wanker by someone or other.

"Well, if you're not, you will be soon because, boy, do we have a lineup for you tonight! We've got dancers and DJs, mimes and musicians, jugglers and gymnasts, artists and acrobats—you name it, we've got it. Each one of tonight's contenders will be competing for the grand prize of ten thousand pounds, which will help the lucky winner get off the streets and start to rebuild their broken life."

A murmur rose from the crowd as people exchanged puzzled looks.

"You know," continued the host, "when they canceled my TV show a few years ago—*Two Short of a Threesome*, I'm sure you all remember it—I ended up living on the streets, and let me tell you, it wasn't easy. I had to do a lot of crazy stuff to survive. Stuff I'm not proud of. But I'd just like to clarify right here and now that despite what certain newspapers reported, I never, *ever* sold my body in exchange for methamphetamine. It's important you all know that. Anyway, what I'm trying to say is that living on the streets is damn hard, as all of tonight's performers know, and so—"

A lanky man with shaggy hair ran onto the stage and delivered a note before scuttling off again.

"Fan mail," said the host. Nobody laughed, which made his own laugh sound somehow worse than the silence.

Removing a pair of glasses from his breast pocket, he put them on and read the note.

"Okay," he said, "so even though a lot of them might *look* homeless, I've just been informed that as far as we know, tonight's performers actually live in houses and some even have proper jobs. Sorry for the confusion there. Maybe forget everything I just told you. Except for the whole selling-my-body thing. Which, let me remind you, was a complete and utter fabrication. Anyway," he said, checking his watch, "the show's about to start, but before we get to the opening act, I think it's time we introduced tonight's celebrity judges!"

Two men and one woman who were sitting at a table in front of the stage appeared on a giant television screen behind the host.

"We have Dave Davidson, otherwise known as Tricky Dicky from Channel Five's hit TV series *Oliver Twisted.*"

A middle-aged man wearing a white shirt, dark glasses, and more fake tan than a Hartlepool hen party waved as the crowd cheered.

"In the middle we have Sarah Buckingham, hard-talking presenter of the award-winning docu-series *Get Off the Dole, You Dirty Scrounger.*"

A slender blond woman in a black suit appeared on the screen. She looked like she'd tortured animals as a child and still thought about it on a regular basis.

"And last, and also least, we have the producer of several popular TV series and the executioner of at least one of them, namely *Two Short of a Threesome*. Ladies and gentlemen, please start slow-clapping for Martin Gould, the man who ruined my life!"

A chorus of boos rolled over the crowd as a bald man in his midfifties appeared on camera.

"Good to see you, Martin," said the host. "Love what you've done with your hair."

Martin forced a stiff-jawed smile like he'd just been puked on by somebody's baby.

"Anyway, without further ado, let me ask you all to give a warm welcome to our first contestant. He's thirty-three, he's from Sheffield, and he's blind, but if you think that's going to stop him from juggling chain saws, then think again! Give it up for Juggling Joe, everybody!"

"What's going on out there?" asked Danny, who'd stayed behind to stretch while Krystal and Will went to watch the opening act with Ivan and Mo. "I thought I heard screaming."

"A guy just tried to kill the judges," said Krystal.

"A blind guy," added Will.

"With a chain saw."

"Four chain saws, actually." Will wiggled four fingers.

"He almost fell off the stage and landed on the judges' table."

"Almost?" said Danny.

"Somebody caught him just in time," said Krystal.

"Shame," said Danny.

Krystal smiled. "You're not getting out of it that easily," she said. "And anyway, if that's the level of competition you're up against, then trust me, you've got absolutely nothing to worry about."

"That's not quite true though, is it?" came a familiar voice from outside the cubicle. Before Krystal had time to choose the right profanity for the occasion, El Magnifico appeared in the doorway, his face sporting several Band-Aids from his recent tussle with Milton. "You *do* have something to worry about, don't you, Danny?" he said.

"Like what?" said Danny.

"I think you know."

Danny thought about this for a moment.

"Global warming?" he said.

"No," said El Magnifico.

"The plague?"

"No."

"Wasps?"

"No."

"You're not scared of wasps?" said Danny. El Magnifico sighed.

Will joined in. "Zombie apocalypse?"

"Forgetting to empty your pockets before doing the laundry?" said Krystal.

"Accidentally making eye contact with somebody on the Tube?" said Danny.

"No!" snapped El Magnifico, growing impatient.

"Getting stuck behind somebody in the supermarket who wants to pay for everything with coupons?" said Krystal.

"Getting maced on your own doorstep?" said Danny.

"Good one," said Krystal.

"It's me, you fools!" yelled El Magnifico. "It's me you've got to worry about! Me! El Magnifico!"

"We *do* worry about you, Kevin," said Krystal, her voice infused with faux concern. "Everybody worries about you. You're precisely the sort of person that society worries about."

"Will, this is the kind of man I'm always telling you not to get into cars with," said Danny.

"That's right, keep laughing," said El Magnifico. "Enjoy yourselves while you can. I can't wait to see how smug you all look when I go home with the prize money and you go back to your pathetic little lives." He turned to Will. "Tell me, *boy*, how does it feel to have an endangered species for a father?"

"How does it feel to have your scrotum ruptured by a size seven stiletto?" said Krystal. She took a step towards El Magnifico.

"Now, now," he said, taking two steps back. "There's no need for violence. I didn't come here to fight. I came here to wish you the best of luck tonight, *actually*."

"How very sportsmanlike of you," said Danny drily.

"I mean, I wouldn't say that if you had a shot at winning, of course, but as you've got more chance of shitting a live pig, I figure there's no harm in saying something generically encouraging. Anyway, I'll leave you to practice. I'm sure you could use it. And remember, it's not the winning that counts. It's watching me win that really matters! Shazam!"

He lobbed one of his smoke bombs into the cubicle and scurried away, unaware that the paper grenade had failed to activate. Everybody was still staring at it when he ran back a few seconds later, scooped up the bomb without saying a word, and ran off again.

"That was Krystal's ex-boyfriend," said Danny, leaning close to Will.

"Seriously?" said Will. Danny nodded. Will burst out laughing.

"What's so funny?" said Krystal, impaling them both with her eyes.

"Nothing," said Danny, struggling to keep a straight face. "Absolutely nothing."

"Ladies and gentlemen," said the host as he reappeared onstage clutching the microphone. "Please give a big round of applause for Elastic Emma!"

Behind him, a young woman in a black leotard unfolded her limbs from the human knot she had somehow managed to turn herself into.

"What did our judges make of that, then?"

The three judges appeared on the huge television screen. Dave gave an enthusiastic thumbs-up. Martin nodded with approval. Sarah shrugged like she'd just been asked how she felt about human rights.

"You know, I wish I was that flexible," said the host, waving to Emma as she walked offstage. "My ex-wife's more flexible than me, and she's been dead for ten years!"

A wave of cautious laughter passed through the crowd.

"It's a joke! It's a joke, don't worry." He paused to allow for a couple of guilt-free chuckles. "She's actually only been dead for *five* years!"

A single cough punctuated the otherwise perfect silence.

"Anyway, on with the show. Next up we have Gerry, short for Geraldine. Gerry likes spending time with her grandchildren and making homemade bird feeders while watching *Nazi Megastructures* on National Geographic. Despite being almost eighty, Gerry loves to dance. She says it keeps her young, and when you see her moves you'll understand why, so put your hands together for Gerry, a.k.a. the break-dancing granny of Gospel Oak!"

Many of the performers had taken to practicing in the corridor, Danny and Will included. They tried their best to avoid the meandering plate spinners, rogue unicyclists, errant acrobats, and pirouetting ballet dancers fighting for space around them while Krystal supervised from the relative safety of the cubicle doorway.

"Looking good, guys," said Tim, a rolled-up cigarette hanging from his lips. "You two a double act now?"

"I was jealous of Milton so I decided to get my own sidekick," said Danny. He ruffled Will's hair.

"Sidekick?" said Will. "*You're* the sidekick."

"Actually, you're both *my* sidekicks," said Krystal.

"And I'm his," said Tim, nodding at Milton. "He's the real brains behind the operation."

"Want to swap him for Danny?" said Krystal.

"I'm not sure he'd fit on my shoulder."

"I also don't look good in sweaters," said Danny.

"Or anything really," said Krystal.

"Hilarious," said Danny.

"Sorry to interrupt," said the portly magician Danny had seen earlier. His face was red like he'd just been Heimliched, and he leaned on Tim to catch his breath. "You haven't seen a rabbit by any chance, have you? He's white, about this big, answers to the name of Derek. He's got long ears and teeth and, well, he's a rabbit."

"Sorry," said Danny.

"Only cats and pandas here," said Krystal.

The man sighed and wiped his face with his colorful string of handkerchiefs. Taking a blast of his asthma inhaler, he ran off in search of Derek.

"Let's hear it for Gerry!" said the host. Gerry waved feebly as two paramedics took her off on a stretcher. "I don't think I've ever seen break-dancing taken so literally before. That wasn't hip-hop. That was hip-POP!"

The audience groaned.

"Get it? Hip-pop? Whatever. Next up we've got a very special musical double act from Peckham. One of them is hairy and never cuts his nails, and the other one is a cat! Put your paws together for Timothy and Milton, everybody!"

A glowing sea of smartphones rose above the crowd as Tim and Milton appeared onstage. Nobody had a free hand to clap with, but the sound of a thousand "awws" made up for the lack of applause.

"Good evening, everybody," said Tim, brushing his hair out of his face. "My name's Tim and this stylish fella right here is my good friend Milton. Say hello, Milton."

Milton meowed into the microphone and the crowd dissolved into mush. Even Sarah smiled briefly.

"We're going to play you a song tonight. I hope you enjoy it. Milton chose it, actually. I'm sure you'll understand why."

He strummed his guitar and the crowd began to sway as the opening bars of "What's New Pussycat?" rang out around the park.

Danny and Will flinched in unison as "Uptown Funk" started blasting from Krystal's handbag.

"All right, you sexy bitch," she said, pressing her phone to her ear. "What? . . . Great. I'll be there in two mins. . . . Okay, babes. Ciao."

"Who was that?" said Danny.

"El Magnifico?" said Will. Danny sniggered as they shared a high five.

"Say that again and I'll high-five your nose," said Krystal. "Grab that bag of face paint and follow me, funny man."

"Where are you going?" said Danny.

"Last-minute prep," said Krystal.

"Last-minute prep? What last-minute prep? You didn't mention anything about last-minute prep!"

"Don't get your fur in a tangle," she said. "Just stick to the plan and leave the rest to me."

"How can I stick to the plan if I don't know what the plan is!"

"You're going to be awesome, don't worry," said Will as he hugged his dad. "Mum would be well proud of you. *I'm* well proud of you. And Krystal is well proud of you too, even though she won't admit it."

"All right, you," said Krystal, gently squeezing Will's shoulders. "Move your arse before you make my mascara run." She looked at

Danny as if she wanted to say something but couldn't find the right words. She nodded at Will. "What he said."

Danny smiled and watched them go before turning his eyes towards the canvas ceiling.

"Looks like it's just the two of us," he said.

The audience erupted as Tim took a bow, careful not to bend too low so Milton wouldn't fall off his shoulder.

"Thanks, everybody," he said. "Say thank you, Milton." Milton meowed into the microphone and everybody melted all over again.

"How about that for a purrrfect purrrformance?" said the host.

The crowd cheered in agreement, warming to the acts if not to the host's crappy jokes.

"I don't know about you, but I'd say it's going to take something pretty spectacular to compete with a performance like that, something really amazing like, I don't know, a *singing dog* or something. But where would we find one of those? Wait a minute," he said, moving his hand to the earpiece he wasn't wearing and pretending to listen intently. "You'll never guess what, folks. That's exactly what we've got for you tonight! Please give a warm welcome to Jack and Daniels!"

The crowd laughed as a giddy Jack Russell galloped across the stage and leapt onto a folding chair positioned beside a short microphone stand. His owner, Daniels, the same elderly gentleman that Danny had seen practicing earlier, shuffled out a moment later, still wearing his white T-shirt with Jack's face printed on it but now also sporting a porkpie hat that looked like it doubled as his dog's favorite chew toy. He grabbed the mic from the taller stand and tapped it a couple of times, then gestured to someone behind the curtain. The Spice Girls started booming from the speakers as Jack, Daniels, and everybody over the age of twenty-five sang the opening verses of "Wannabe."

The following act was a seven-man Japanese dance troupe who ran around the stage dressed like robots, sometimes chasing each other and sometimes being chased by a large squid monster with flashing red eyes. Next up was a Belarussian strongwoman who twisted steel rods into various shapes as if she were twisting balloons before throwing them into the crowd as gifts, a charitable gesture that was sadly offset by several minor head injuries. A unicyclist was followed by a ballet dancer, an escape artist, a sword swallower, and a mime. A bunch of Mexican acrobats called Circ du Olé spent five minutes swinging from the rigging while dressed in comically oversize sombreros, and a snake charmer from Wigan almost fainted while furiously playing his pungi until finally giving up when Fred, his Indian cobra, made it absolutely clear that he was not going to leave his basket. A teenage rap duo and a martial artist with a wooden leg came next, and they were followed by the chicken man, the nut juggler, the one-man band, and the human statue, until, with just two performances remaining, it was El Magnifico's turn to take the stage.

The host came out to introduce him, marching from behind the curtain with his crumpled blazer flapping open. He looked like he'd been drinking.

"Is everybody having fun?" he shouted.

The crowd responded mostly in the affirmative.

"I can't hear you!" he yelled, a little belligerently, as if he actually hadn't heard them and hated asking the same question twice. "I said, is everybody having fun!"

People murmured in a way that suggested they'd be having a lot more fun if he stopped asking them if they were having fun.

"That's better! You know, I'm glad I'm not the one judging tonight's competition, because I'd give everybody first prize. Who do *you* think deserves to win?"

Everybody answered at the same time, their words fusing

together to create the impression that the favorite act of the night so far was a Mexican clown charmer.

"Well, before you decide who should be crowned Mr. or Mrs. Street Performer of the Year, hold your horses because the show isn't over yet! We've got two performances left, and this next one is guaranteed to blow your mind. We don't know his real name and we don't know where he comes from. All we know for sure is that he's going to show you something that may just make you question everything you've ever known. Without further ado, let's hear it for El Magnifico!"

Cheers rose up from the crowd before petering out as the stage remained empty. A nervous silence descended, while people shuffled their feet and waited for something to happen. Even the host seemed anxious, repeatedly checking his watch and exchanging uneasy glances with the organizers. He was just about to walk back out to deliver some impromptu and potentially tragic fiction about the fate of El Magnifico when every light in the venue went out. Only when a resonant voice emerged from the gloomy enclave of the stage did the audience realize that the power cut was deliberate.

"Harry Houdini once said, 'My brain is the key that sets my mind free,'" intoned El Magnifico. He let the words fade out for effect before a single spotlight revealed his presence in the middle of the stage.

A hushed excitement rippled through the crowd.

"But even Houdini couldn't do what a ten-year-old Italian boy named Benedetto Supino did. While sitting in a dentist's waiting room in 1982, little Benedetto set fire to a comic book he was reading. He didn't use matches or a cigarette lighter. No. Benedetto Supino used his mind! And tonight, in front of your very eyes, via the power of pyrokinesis, I'm about to do the same."

A second spotlight beamed down in front of the stage. Sitting in the middle of the bright white light was a podium, and sitting on the podium was a stuffed panda toy.

"Before I begin, I'd like to warn you not to try this at home," said El Magnifico. "Due to the incredible amount of energy required for this particular feat, the human head has been known to explode while attempting pyrokinesis, so not only am I about to show you something that you've never seen before, nor ever will again, I'm also risking my very life in the process."

At the mention of a possible exploding head, everybody shuffled forward. Beethoven's Seventh Symphony began to drift over the crowd.

Placing his fingertips on his temples, El Magnifico flexed his jaw and stared at the bear. His face began to tremble like a kettle coming to boil and he frowned so deeply that his eyebrows overlapped. A high-pitched sound emerged from his lips that caused Jack the dog to start barking backstage; but sixty seconds later the bear looked no closer to being immolated, something it seemed almost smug about as it grinned at the crowd from the giant television screen. Nobody saw Derek the rabbit lollop along the hem of the stage and disappear beneath the judges' table, their eyes fixed firmly on the quivering magician who had now turned such a violent shade of red that several paramedics had gathered nearby.

As one minute became two and two became three, boos rose up from the front of the crowd and rapidly spread until everybody was jeering, including the judges. Undeterred by the growing dissent, El Magnifico redoubled his efforts. He focused intently on the bear, staring at the toy as if he didn't just want to cremate it but will it out of existence entirely. Just when he was about to be dragged away by the security staff who had appeared on either side of the stage, a huge gasp went up from the crowd as the judges' table burst into flames. Even El Magnifico looked shocked as Martin slapped at his flaming trousers while Sarah, despite not actually being on fire, screamed and rolled around on the ground until the fire crew proceeded to blast them both with fire extinguishers. Realizing he'd actually managed to set something on fire—not the *right* thing, for the panda hadn't even broken a sweat, but *something*—El Magnifico's body started to tremble again, this time with fits of laughter. He didn't notice the fire crew dragging the charred corpse of Derek the rabbit out from under the table, the electrical cable he'd unwisely chosen to snack on still firmly clamped between his blackened jaws.

The host lingered on the side of the stage farthest from El Magnifico.

"El Magnifico, ladies and gentlemen," he said a little unsurely.

The magician took an extravagant bow before swaggering off the stage like someone who was already ten grand richer.

"Well, this competition's really heating up!" said the host.

Nobody laughed or even bothered to groan, far too shocked by what they'd just witnessed.

"There's only one more act to go before the judges decide on the winner," he continued. "Will it be better than the rest? We're about to find out, so please raise the proverbial roof for our final performance. We've had enough chaos tonight, so now it's time for a little . . . Pandamonium!"

CHAPTER 33

Danny peered out from behind the curtain, his pulse galloping and his heart pounding so fiercely that every nervous beat made the mangy fibers of his black-and-white chest fur twitch.

A single strobe flickered in time to the music that began to pump from the speakers, exposing the stage every four or five seconds with a brilliant burst of bright white light. Down in the crowd, Ivan cheered as if Ukraine had just won Eurovision and Mo whistled so loudly that his hearing aid malfunctioned. Others watched with a numb sense of duty, convinced that whatever was about to happen couldn't trump the experience of seeing the judges flambéed with mind power.

The longer the stage remained empty, the more impatient the crowd became, and just when the boos were about to ring out, a figure emerged from the darkness, revealed by the flash of the strobe. The shape returned a few seconds later in another blinding flicker of white. A third and a fourth flash quickly followed as the beat grew more intense, and when the intro reached its peak in a drumroll of sound that came to a halt with an eerie silence, the stage remained illuminated just long enough for the crowd to get a proper glimpse of the mysterious figure in the middle of it. Some people whispered

"Rat." Others said "Raccoon." "Badger" was also mentioned. Several variations of "What the fuck is that?" could be heard in different sections of the crowd. A few children cried. The lead singer of God Knows What muttered something about the devil. Nobody knew what they were looking at, but nobody could look away.

Danny stared ahead of him, frozen with fear. He felt like a first-time skydiver waiting to jump, and the longer he stood there, the more time he had to contemplate just how many things could go wrong, until he couldn't for the life of him remember a time when any of this seemed like a good idea.

He tried to move, but his body wouldn't listen. He couldn't even *feel* his arms and legs, never mind encourage them to dance. He momentarily thought he might be having a heart attack, and when he realized he wasn't he thought about faking one, dropping to the floor and pretending to be dead until a stretcher arrived to carry him away. He squinted through the lights at the faceless crowd in the hope that somewhere, out there, a rogue assassin with a bullet to spare would put him out of his misery; but the longer he stared, the more he saw that the crowd wasn't faceless at all. He could see somebody, vaguely at first but then as clearly as if they were standing right in front of him. He closed his eyes and opened them again, but Liz was still there, watching him with the faintest of smiles, a smile that told him that she was okay, and that he was okay, and that everything was going to be okay just as long as he got his shit together, right now, and danced like he'd never danced before.

So that's what he did.

He didn't even know he was moving at first, his body having wrested control from his mind like a passenger taking the wheel from a dead man, and it wasn't until he heard the crowd roar that he realized he was dancing, the sound surging through him like he'd just been shot in the arse with adrenaline. He cut through the air like the sword of Zorro and he spun with the speed of a weathervane in a hurricane, his mind and body having reconciled their differences and now working in perfect harmony to deliver a show that was worthy of all the hard work and patience (mostly Krystal's) and the sweat and tears (mostly Danny's) that had gone into making it.

He didn't miss a single beat, even during the feverish crescendo that often tripped him up in rehearsals, and he crackled with such frenetic energy that he had to consciously temper his moves for fear of overplaying them. He didn't even want to stop at the inter-lude, something he often couldn't wait to do, the precious handful of seconds giving him just enough time to catch his breath and say a quick prayer for the strength to continue before staggering back into the fray. This time the interlude felt more like a nuisance than a respite, interrupting his flow when he was just warming up and forcing him to wait for the second phase to start, a phase also known as Krystal's Brilliant Idea.

Taking position in the middle of the stage, where he'd hoped to die a violent death less than a minute previously, Danny quietly counted down as the music climbed towards the next heavy beat drop. The lights went out and the strobe began to pulse the way it had during the intro, blinding flashes that lit up the eyes of the crowd, all of them wide with excitement as they stared at Danny alone on the stage and wondered what he was about to do next until, suddenly, impossibly, he wasn't alone anymore.

"Ready?" he said to the miniature, shabbier version of himself that had somehow appeared beside him in the split second between the flickering lights. Krystal had promised that the fabric they'd torn from Fanny's couches had been thoroughly fumigated prior to running it through her sewing machine, but as he watched Will scratching himself like a guard dog at a flea farm, Danny once again wondered how he'd ever let Krystal talk them into this.

"Ready!" yelled Will as the beat kicked in. They started to dance in sync, shimmying, shaking, spinning, and strutting across the stage while the audience cheered so loudly that the music had to fight to be heard. They hadn't had much time to practice. Danny's only directive from Krystal had been to keep on dancing when Will appeared, and focus on nothing but his own performance. The real onus was on Will, whose job was to make sure that everything he did corresponded with everything his dad was doing. If his dad sped up, he sped up. If his dad slowed down, he slowed down. If his dad fell headfirst into the crowd, he fell headfirst with him in the hope of making it look like an orchestrated stage-dive. Whatever it took to maintain the synchronicity. The setup was far from ideal, and the plan was nerve-rackingly dependent on Danny following the script; but despite everybody's concerns, his own included, Danny moved when he should have moved, he hit every beat on the list, his timing was down to the millisecond, and he didn't even go near the crowd, never mind fall off the stage. Even if he had, everybody was far too busy dancing to notice.

"Where's Krystal?" he yelled when the song arrived at its second, final interlude. He hadn't seen her since she'd abandoned him backstage and he was concerned that something bad had happened, not to her but to whoever was stupid enough to mess with her.

"What?" shouted Will as he furiously scratched his arse.

"Krystal!" he said, breathing heavily and sweating profusely. "Where is she!"

"Behind you!" said Will.

Fully expecting to find Krystal grinning at him through the curtains, possibly giving him the finger or two, Danny turned around to find several women lined up on the stage behind him. Their eyes, lips, and noses were colored black and the rest of their faces were painted white. Each wore a matching panda costume complete with a headband sporting stubby little panda ears, and Danny had no idea which one was Krystal until the panda in the middle blew a bubblegum bubble.

"I'll explain later," she yelled, unable to see his face but sensing his confusion. "Just stick to the plan and leave the rest to us."

Danny nodded dumbly, as if he'd just been told his cat had died when he didn't even own a cat.

Turning to face the crowd, he took a deep breath and held it while he readied himself for the grand finale. His lungs burned, his muscles ached with every movement, his limbs felt heavier than waterlogged timber, and his inner thigh had been rubbed so raw that it probably looked like salami, but when the beat dropped for the third and final time, Danny dug as deep as he could and threw whatever he dredged from the bottom of his soul into that last sixty seconds. His energy stores had been thoroughly ransacked, so he borrowed the energy that came from the crowd, their whistles and cheers inspiring him to dance faster, to double down and push through no matter how much his heart threatened to pop from his chest like a fat man's shirt button. He gave that performance everything he had, taking his body to its limits and beyond, so by the time the music came to an end, the climax galloping headlong into a sudden wall of silence, he once again felt the urge to clutch his chest and drop to his knees, only this time he wouldn't be faking it.

He looked at Will beside him and the two of them shared a weary thumbs-up. He looked at Krystal behind him, who nodded with approval as she posed with the others. He looked at the crowd in front of him, their faces now illuminated by the giant spotlights that shone down upon them from every corner of the venue. And then, to chants of "Pandamonium!" that grew progressively louder until Danny could feel the sound in his bones, the women took a step forward, Danny and Will took a step back, each of them linked hands with the person beside them and bowed as applause exploded around them. Only one person remained silent and that was El Magnifico, who looked like he was trying to burn the entire venue to the ground.

"Pandamonium, everybody!" said the host as Danny, Will, Krystal, and her entourage took another bow and disappeared behind the curtain.

"I told you they'd love him!" said Krystal backstage, jabbing a fake fingernail at Will. "Didn't I tell you they'd love him?"

"Whatever's still living inside that fabric clearly loves him," said Danny as Will wriggled out of his costume as if it were on fire. "I thought you said that thing was safe!"

"No, I said it was *clean*," said Krystal. "I have no idea what they cleaned it *with*."

"Why are my arms green?" said Will, holding up his forearms, which were now the color of Granny Smiths.

"Good question, Will. Why is he green, Krystal? Look at his arms! Look at his face!"

"My face too?" said Will, his panic rising.

"Oh yeah, the guy told me that might happen," said Krystal. "Don't worry, he said it'll probably go away in a couple of weeks."

"Probably!" said Danny.

"Weeks!" said Will.

"Look, it did the job, didn't it? And anyway, it's a small price to pay for bagging ten grand."

"If we win," said Danny.

"Are you kidding me? You totally smashed it out there."

"You weren't so bad yourself," said Danny. "Nice touch with the backup dancers, by the way. I don't know how you pulled that off, but, well, thanks."

"Don't thank me," said Krystal. "Thank Fanny."

"Fanny?"

"She told the girls they could have the night off if they came down here to help you."

"Really?" said Danny, taken aback. "She never struck me as the charitable type."

"She is when she wants something," she said, giving Danny a smile that instantly made him uncomfortable.

"And . . . what does she want, exactly?"

Krystal's smile widened. "You," she said.

Danny laughed and waited for Krystal to join in. She didn't. He stopped laughing and cleared his throat.

"Well, that's very flattering, really, and maybe if I was, like, I don't know, three hundred years older, but probably not even then, to be honest—"

"Not like that, you muppet. Fanny's not *that* desperate. She's thinking of starting a weekly ladies' night down at the club, but she doesn't have any male dancers yet, so—"

"No," said Danny, suddenly realizing where the conversation was going. "Absolutely not."

"Why not? You just danced in front of a massive crowd of people, I'm sure you can shake it for a few hammered housewives."

"I had clothes on! There's a big difference!"

"You won't have to be naked, Danny, don't worry. Even fish wouldn't want to see your maggot."

"Really?"

"Well, okay, maybe a really hungry fish might, but—"

"No, I mean I won't have to be naked?"

"No," said Krystal. "Well, not like *naked* naked. Just, you know . . . more naked than not naked."

"You know, as tempting as that sounds, I think I'm going to pass on this one. Will would never forgive me, would you, mate?"

"Doesn't bother me," said Will with a shrug. Danny scowled at him.

"It's two hundred fifty pounds a night, Dan, in the hand. Or thong. Whatever you prefer."

"I don't care," said Danny. "Wait—how much?"

"You heard me. A grand a month for four days' work. Plus tips.

Think about it," she said before going to get changed. Will walked off in the opposite direction.

"Where are you going?" said Danny.

"To wash this off!" said Will.

"I thought green was your favorite color," said Danny, trying to keep a straight face.

"Not anymore," he grumbled on his way to the bathrooms.

Inside one of the Portaloos, Will scrubbed himself with handfuls of soap, but the toxic tint refused to budge. As if things couldn't get any worse, he burst out of the toilet and bumped right into the one person he really didn't want to see at that moment.

"Malooley," said Mark, turning Will's very own name into something that sounded like a threat. He looked at Will's face and arms and frowned. "Why the fuck are you green?"

"Long story," said Will.

"It's an improvement."

Will nodded but said nothing, already tired of this conversation.

"Where's your boyfriend?" said Mark, looking around for Mo.

"Where're your Minions?" said Will, also scanning the crowd.

Mark glared at him for a long moment, a moment that Will was sure would end in pain, which was why he was so surprised when Mark cracked a smile, a far from pleasant sight but still more welcome than the alternative.

"Their mums won't let them stay out this late," he said.

"Well, it *is* a school night," said Will. They shared a brief laugh.

"What you got there?" said Mark, pointing at the panda suit spilling out of Will's bag.

"Nothing," said Will. He tried to hide the bag behind his back, but Mark had already spotted the black-and-white mask.

"Wait, was that you onstage just now? Pandamania or whatever?"

"Pandamonium."

"Fucking hell, that *was* you!" he said.

Will braced himself for the barrage of insults he knew was coming. Dancing is gay. Pandas are gay. Dancing pandas are gay. Something along those lines.

"You were fucking..." Mark groped around for the right word. "...epic!"

"What?" said Will, caught off guard. He wondered if Mark knew what *epic* meant.

"Seriously, you lot were sick, mate, best thing I've seen all night."

"Thanks," said Will hesitantly.

"Where'd you learn to dance like that?"

"My mum."

"Nice. I wish my mum had skills like that. Some days she doesn't even get out of bed." He laughed, but the sound was hollow.

"I could teach you," said Will. "If you want, I mean."

"Me? Dancing?" Mark laughed. "Good one. Any chance you could introduce me to them girls you were dancing with, though?" Mark played with his fringe, as if scruffy hair was the only thing standing between him and the cover of *Esquire*.

Will smiled. "I'll see what I can do," he said.

"Nice one. Anyway, I better bounce. Good luck tonight, yeah?"

"Fingers crossed," said Will.

Mark turned to leave. "Oh, and Will?" he said over his shoulder.

"Yep?"

"Tell anybody about this conversation and you're dead," he said with pseudo-seriousness. "Got that?"

"Got it," said Will.

"Good. See you later, loser."

Will watched him go, wondering if he'd gained a friend or simply lost an enemy.

*　　　*　　　*

Danny was standing near the front of the stage with Krystal, Mo, and Ivan, who stood about half a meter higher than anybody else in the crowd.

"Here's the star of the show!" said Mo. He went to hug Will but froze before contact. "Wait, why is your face—"

"Ask Krystal," said Will. Mo looked at Krystal. Krystal looked elsewhere.

"You're just in time," said Danny as the host reappeared onstage.

"Well, everybody, it's the moment you've all been waiting for!" the host said.

The crowd cheered.

"No, not the part where I stop talking—"

A chorus of boos rose up from the crowd.

"Don't worry, that'll happen in a minute—"

The boos instantly turned into cheers again.

"But before I go home, drink a bottle of scotch, and maybe put a gun in my mouth—"

The crowd seemed undecided on this one.

"Only a joke. I don't have a gun. And even if I did, I wouldn't be using it on myself, would I, Martin?"

Martin flashed him the finger before pretending to scratch his nose with it when he saw himself on the video screen.

"Anyway, it's time to reveal tonight's winner! As you all know, only one act will walk away with the grand prize of ten thousand pounds, so what do you say, judges? Have you made your decision?"

The judges looked at each other and nodded. An envelope was carried onto the stage.

The host put on his glasses and squinted at the note inside.

"Okay, let's get down to business," he said. "Coming in at third place we have . . . Tim and Milton!"

Everyone applauded as Tim and Milton appeared on the video screen. Tim waved and smiled for the camera.

"Don't look at me like that," he whispered as Milton glowered at him from his shoulder. "You chose the song, not me."

The host looked at the note and adjusted his glasses.

"Coming in at second place we have . . . drumroll, please . . . Pandamonium!"

Will sighed. Mo swore quietly. Krystal swore loudly, so loudly that anybody near her shuffled back a few inches, but Danny said nothing as he stared at the stage, his eyes still wide with anticipation as if he hadn't heard the host and was still eagerly awaiting the announcement.

"You okay?" said Ivan.

Danny didn't respond, paralyzed by disbelief.

"Well done to the dancing pandas. You might not have won the grand prize, but you *have* won the next best thing, which is four free tickets to my upcoming stand-up tour!"

A number of the other contestants suddenly looked relieved that they hadn't finished second.

"And so it is my great honor to reveal that the winner of this year's Battle of the Street Performers is . . . and it comes as no surprise because, let's be honest, the performance was pretty mind-blowing . . ."

El Magnifico grinned as he readied himself for victory.

"Let's hear it for Jack and Daniels, everybody!"

The magician's smile slipped from his face as the old man and his Jack Russell took to the stage to collect their prize.

"No," he muttered. "That's not right. *I* won. Me."

"This is fucking bullshit!" said Krystal. "Pardon my French, Will."

"This *is* fucking bullshit," said Will. "Pardon my French, Dad."

Danny remained silent, unable to find even the right facial expression to convey his emotions, never mind the right words.

"Congratulations," said the host, handing Daniels an absurdly oversize check. Before the old man could grab it, however, El Magnifico burst from the crowd and wrested the prize from the host.

"I won!" he yelled, his eyes wild and his mouth foaming slightly. "Me! Not you! I set something on fire with my mind! WITH MY FUCKING MIND!"

One of the guards rushed onto the stage and took out El Magnifico with a crunching tackle, while another guard prised the check from his fingers. Even Jack got involved, clamping his jaws around El Magnifico's robe and refusing to let go.

"Get off me!" yelled El Magnifico. "Get off before I burn this fucking place to the ground! I have powers. Didn't you see? I HAVE POWERS!"

The guards dragged the magician away to the sounds of cheers and laughter. More laughter followed when Jack came trotting back with a piece of El Magnifico's robe in his mouth.

"I believe this is yours," said the host, picking up the slightly crumpled check and handing it to Daniels. "Give it up for this year's winners, ladies and gentlemen!"

A well-groomed young man appeared onstage and presented the performers with a trophy, which Daniels held above his head to the deafening chants of "Encore!"

"It sounds like people want to hear you sing again, Jack," said the host, bending down to address the terrier. "What do you reckon?"

Jack barked once, and everybody cheered.

"I think that's a yes!" said the host. "Take it away, guys!"

"Let's get out of here," said Danny.

CHAPTER 34

Ivan set four drinks on the table, the glasses like thimbles in his meaty palms.

"Cheers!" he said, raising his pint in toast.

Krystal took her triple gin and tonic. Will took Danny's pint. Danny took the glass of Coke and swapped it with Will. Everybody clinked glasses, but only Ivan's clink had any enthusiasm behind it.

"Sorry, Ivan, but what are we celebrating exactly?" said Danny.

"Life!" said Ivan, sweeping his hand around the Cross-Eyed Goat, as if the dingy pub and its miscreant clientele were the perfect reasons to celebrate existence.

"Yeah, well, life doesn't seem like something worth celebrating right now," said Danny.

"Fine. You want good news? I give you good news. Remember Viktor? Asshole from building site? He go back to Russia after somebody put his head in toilet."

"Why did they do that?" said Will.

"Must have messed with the wrong people," said Krystal, exchanging a smile with Danny.

"See," said Ivan. "You smile. Life is good."

"Not when Reg finds me," said Danny. He sighed and took a contemplative sip of his pint. "I just don't get it. I really thought we'd won tonight."

"We *did* win," said Krystal. "I mean, we didn't 'win' win, obviously, but we were miles better than anybody else. The crowd loved you."

"The crowd loved *you*," corrected Danny.

"The crowd loved *us*," she said.

"Then how did we get beaten by a singing dog?" said Danny.

Will shrugged. "At least we didn't get beaten by El Magnifico," he said.

"Exactly," said Krystal. "That's something to celebrate, right?"

"That's true," said Danny, grinning at the thought of a rabid El Magnifico getting slammed to the floor by security staff.

"You know that guy was Krystal's ex-boyfriend, right?" said Will to Ivan. Ivan looked at Krystal. Krystal looked at Will. Will looked at Danny. Danny laughed nervously. He didn't hear Reg walk through the door behind him.

"Malooley!" he yelled. Everybody flinched, apart from Ivan.

"Great," said Danny, not even bothering to turn around as Reg made his way towards him. "I can't even enjoy losing in peace."

"Hello, Daniel," said Reg. Mr. Dent loitered by the door, presumably in case any of them felt like running.

"Hello, Reg."

"What's the occasion?" he said, eyeballing the table, especially Krystal.

"We're having a wake," said Danny.

"Oh yeah? For who?" said Reg.

Danny downed his pint while he still could. "For me," he said.

"I assume that means you ain't got my money."

"You assume correctly, Reg," said Danny. "I do not have your money." He sounded more weary than worried, like a death row inmate who was tired of waiting and just wanted to get the show on the road.

"Charlie!" yelled Reg. "Bolt that door, would you."

The sound of scraping chairs and frantic footsteps filled the room as everybody rushed to get out before Charlie locked them in. He bolted the top and bottom of the door and barred the middle with a plank. Then, turning off the security cameras with a casualness that suggested this wasn't the first time he'd done this, he grabbed a bag of Scampi Fries and disappeared upstairs.

"Well, don't just fucking stand there, Dent," said Reg. "Get a fucking move on. *EastEnders* is on soon."

Lurching into motion like a ten-ton truck, Mr. Dent lumbered across the room as Danny got to his feet, not because he had any kind of plan but because standing seemed more dignified than meeting his fate while quaking on a barstool. He felt Ivan's hand on his shoulder and guessed it to be an act of solace until the Ukrainian shoved him back into his seat before leaping up with surprising speed and striding towards Mr. Dent. The two men lunged the moment they were close enough, clamping their giant arms around each other and refusing to let go. The longer they stood there, the more it became apparent that the men weren't grappling but hugging, holding one another in a firm embrace that would have crushed most humans to a pulp.

Everybody looked confused. Especially Reg. They were even more perplexed when Mr. Dent started speaking. In Ukrainian. The two of them chatted like long-lost friends as Ivan led Dent towards Danny, who instinctively shuffled back a few steps to keep himself out of denting distance.

"Danny, I like you to meet Dmitri," said Ivan, roughly patting Mr. Dent on the back.

"You two know each other?" said Danny.

"Of course! He is my brother's boy!"

"Dent's your nephew?"

"Yes. Nephew. Last time I see him, he is this big. Like grasshopper." Ivan indicated a height that was still taller than most tall men.

"Yeah, well, this 'grasshopper' almost broke my legs a week ago."

Ivan frowned at Mr. Dent while Reg looked on in disbelief.

"What is this! A family fucking reunion? Stop fucking about, Dent!"

Mr. Dent buried his hands in his pockets and stared at his shoes like a naughty schoolboy while Ivan yelled at him in Ukrainian.

"Tell him you're sorry," he said, nodding at Danny. Mr. Dent mumbled something, and Ivan slapped him around the head. "In English!"

"Sorry," said Mr. Dent.

"Is okay?" said Ivan.

"I guess," Danny said.

"Shake," said Ivan, grabbing their hands and forcing their palms together like a teacher resolving a playground scuffle.

"Sweet fucking baby Jesus, would you stop talking and start breaking stuff already!" said Reg.

Ivan said something to Mr. Dent, who nodded and looked at Reg.

"What the fuck are you up to now!" said Reg as Mr. Dent lumbered towards him. "You're going the wrong way, you stupid sack of—"

Dent grabbed him by the throat before he could finish his sentence.

Ivan looked at Danny. "What you want to break first?" he said. Reg's crutches clattered to the floor as he feebly grappled with the hand around his windpipe. "Arms? I think arms."

"Hold up!" croaked Reg, his face turning redder than a boxer's urine sample. "Let's talk about this!"

"Or maybe his legs again," said Ivan. "What do you think?"

"Wait!" said Reg. "Danny! Forget about what you owe! We're square! And I'll reduce your rent by . . . twenty percent!"

"You mean the twenty percent you already raised it by?" said Danny.

Ivan nodded at Dent, who squeezed a little tighter.

"Fifty!" squeaked Reg. "Fifty percent!"

"That's very reasonable, Reg, but you know what? I think we're going to move." Danny looked at Will. "If that's okay with you?" Will nodded.

"Can we do the fingers?" said Ivan, determined not to leave without breaking something. "Just the fingers."

"Let him go," said Danny.

"You sure?" said Ivan. Danny nodded. Ivan shrugged and looked at Dent. He said something in Ukrainian and gestured towards the door.

Still clutching Reg's throat, Mr. Dent unbolted the door with his free hand.

"Don't forget those," said Danny, pointing to Reg's crutches. Mr. Dent scooped them up and carried them out along with their semi-conscious owner.

"Thanks, Ivan," said Danny as they both sat down. "I think you just saved my life."

"No problem," said Ivan. "Now we are even."

Danny frowned as if he wasn't quite sure what he'd just heard. "Even for what?" he said, a smile slowly appearing on his face.

"What?" said Ivan, suddenly preoccupied with trying to remove an invisible stain from the table with his fingernail.

"You just said, 'Now we're even.'"

"When?"

"Just now!"

"No, I didn't."

"Yes, you did," said Danny.

"You spend too much time inside costume," said Ivan. "Is bad for hearing." He tapped his ears for emphasis.

"Well, anyway, thanks," Danny said.

Ivan was dismissive now. "I just wish I knew Dmitri he worked for your landlord before, so you don't have to make such idiot of yourself."

Danny looked at Krystal and Will, who were busy strangling each other while playfully reenacting Mr. Dent's chokehold on Reg. He watched them talking and laughing and winked when Will caught his eye across the table.

"Actually, I think everything worked out pretty well in the end." He held up his glass. "Cheers."

"To dancing rats!" said Ivan.

"We're pandas!" said Will.

"Yeah!" said Krystal. "We're pandas!"

"To the pandas!" said Danny.

"Whatever," said Ivan, raising his glass. "Cheers!"

EPILOGUE

The snow was barely half an inch thick, just enough to wreak absolute havoc on every level across the entire Southeast. Flights were grounded or redirected, trains were canceled or delayed, tea bags and toilet paper were flying off the shelves as people stockpiled for the next ice age, and everybody was asking the same question they always asked whenever the weather did precisely what it was supposed to do in England in December.

"Why can't we be more like Sweden?" said the old man opposite Danny as he ran his hand through the condensation and stared at the traffic from the bus window. "You don't see the Swedes getting their underpants in a tangle every time it snows."

"What are you talking about underpants for?" said his wife, who sat beside him.

"I'm not talking about underpants, Edna. I'm talking about the bloody state of this country."

"Well, it's hardly the Swedes' fault, is it?"

"I'm not blaming the Swedes!"

The old man looked at Danny and shook his head. Danny smiled.

"Is this about Oliver?" said the man's wife.

"What?"

"Sarah's husband."

"I know who Oliver is. Why are we talking about Oliver?"

"You never did like him much."

"What's that got to do with anything?"

"He's Swedish."

"He's from Burnley!"

"Not with a name like Gustavsson, he's not. You never did like our Sarah becoming a Gustavsson."

"I'm not talking about Oliver!"

"Then what *are* you talking about?"

The man was about to reply when the bus slowed down and his wife got up and made for the door. He looked at Danny, noted his wedding ring, and gave him a weary smile.

"You've got all of this to look forward to," he said as he stood and followed his wife off the bus. Danny forced a smile and looked down at his ring. He plucked a strand of cotton from it and watched it float to the floor.

Will gave him a gentle nudge in the ribs. "It's okay," he said. "You can always rely on me to make your life difficult."

Danny nudged him back.

"Bring it on," he said.

Despite the bitter cold, the cemetery was busier than usual. Christmas was just around the corner and many of the graves had been adorned with candles, wreaths, and festive red ribbons.

Danny paused beside Liz's headstone and gently wiped the frost from her name.

"There you are," he said. "Hello, beautiful."

Will shuffled up behind him.

"She would have loved all this, wouldn't she?" said Danny as he looked around at the snow. Will nodded but said nothing.

They both stood in silence beneath the dull white sky and stared at the untouched snow that lay across her grave like a blanket.

"You want to say something to your mum?" said Danny. "Wish her Merry Christmas or something?"

Will thought about this for a moment.

"Dad's a stripper," he said. An old lady tending to a nearby grave looked up and scowled at Danny.

"Dad is *not* a stripper," he said, loud enough for their neighbor to hear.

"He is," whispered Will to his mum.

"Don't listen to him, Liz," said Danny. "And don't encourage him either. I know what you're like. I'm not a stripper. I'm a stage performer."

"A stage performer who takes his clothes off."

"A stage performer who makes enough money for us to live in that nice new flat with that nice new bedroom of yours."

"You'd really like it, Mum," said Will. "It's got loads of space to practice."

"We've been working on a new panda routine for the park. Will's been dancing with me on the weekends."

"I have a proper costume now."

"You should see him, Liz, the crowds love him. Especially the girls."

"Whatever," said Will, his cold cheeks turning another shade redder. "Can we show Mum the new routine?"

"What, here?" said Danny, looking around. Will nodded excitedly. "I don't think that's a good idea."

"Why not?"

"Because dancing in a graveyard is just . . . it's not an okay thing to do."

"Mum wouldn't mind."

"Yeah, well, other people might," said Danny, glancing over at the old lady, who was still eyeing him suspiciously. "Save it for Gran and Granddad."

"They're coming to visit next week," said Will, filling Liz in on the news.

"Your dad called me out of the blue and asked if we could do Christmas together," said Danny. "I know, I couldn't believe it either."

"They talked for ages. They apologized and everything."

"*He* apologized. And we didn't talk for ages. We talked for a few minutes."

"You still talked, though."

"Yeah," said Danny. He smiled. "We talked. Speaking of which, tell your mum what Mr. Coleman said about you at parents' evening last week?"

"He said that I'm the nicest, most polite, and most respectful student he's ever had."

"Not that bit. The other bit."

"He said I talk too much in class."

"You hear that, Liz? He talks too much in class! How great is that?"

"I don't think he meant it as a compliment," said Will as Danny proudly ruffled his hair.

"He's also made a new friend, haven't you, mate? What's his name? Matt?"

"Mark Robson," said Will. "He's the toughest kid in school."

"Will promised to introduce him to Krystal, so now Mark protects him with his life."

"And Mo. A kid made a joke about his hearing aids the other day, so Mark wedgied him so badly that he had to go to the nurse's office."

"When are we supposed to be meeting Mo anyway?" said Danny, checking the time.

"One o'clock at the ice rink. Krystal's already there."

"Then we better get our skates on," said Danny. Will stared at him blankly. "Get it? Skates on? It was a joke." Danny grinned.

Will shook his head. "See what I have to put up with, Mum?"

"Come on, let's make a move." Danny touched his lips and placed his hand on the headstone. "Merry Christmas, Liz. Love you."

He plucked the withered flowers from the grave and went to put them in the bin, his shoulders instantly dropping the moment his back was turned. Will watched him walk away and listened to the lonely crunch of snow that followed Danny down the path.

"Don't worry, Mum," he said. "I'll look after him."

END

ACKNOWLEDGMENTS

I would like to express my sincere gratitude to the following people:

Joanna Swainson, my brilliant agent. Thank you for your hard work, your patience, and, above all, for giving me a chance. Also, thank you to Thérèse Coen, Nicole Etherington, and the rest of the team at Hardman & Swainson. I feel very lucky and immensely proud to have you in my corner.

Kara Watson, my editor at Scribner, and Katie Brown, my editor at Trapeze/Orion. This book is so much better because of you. Also, thank you to Joal Hetherington, Sarah Fortune, and Marc Simonsson.

Greg Lovell, a great friend and a great writer, thank you for everything; Richard Skinner and the Faber Academy, for giving me the confidence to take my writing seriously; Kim and Joachim Leckscheidt, for being such wonderful company; Justas Ramanauskas, for all the beers and the brainstorming; Adam and Barbara Libove, for reading my early work; and Sophia Syed, Simon Mount, Aris Macos, and Max Doyle, for all of your encouragement over the years.

Jarrod Gould-Bourn, for making sure I never ran out of Marmite during the writing of this book; Sarra Szmit, for always reading whatever I write; my nephews, Thomas and Harrison Szmit—Tom, thanks for offering to sell signed copies of my book in exchange for all of the proceeds, and Harry, thanks for reading the earlier draft of this novel, and sorry it had so much swearing in it; Bill and Stella Valentino, for making me feel like I'd succeeded long before I actually did; and Linda and Phillip Gould-Bourn, for being with me for every step of this long journey.

And finally, thank you to my wife, Vanessa Valentino, otherwise known as my Secret Weapon (although not so secret now, I guess). Writing is hard at the best of times, but it would be so much harder without you. Thank you for your tireless proofreading, for always laughing at my crappy jokes, for putting me back on that horse whenever I fell (or threw myself) off, for reminding me to shave whenever I was in the midst of one of my feral writing episodes, for sharing the burden of my countless rejections, and for always believing in me, even on those days when I no longer believed in myself.